PICKING UP THE SLACK

Law, Institutions, and Canadian Climate Policy

Canada has over-promised and under-delivered on climate change, setting weak goals and allowing carve-outs, exceptions, and exemptions to undermine its climate policies. Why, in an era when climate change is front of mind for so many people, have we failed to make progress? This question has been the source of heated debate across the political spectrum. In *Picking Up the Slack*, Andrew Green draws together different perspectives on the challenge facing Canada to offer an accessible account of the ideas and institutions that have impeded climate change action.

Picking Up the Slack embraces the complexity of the problem, showing that its sources lie deep in Canada's institutional arrangements – pointing to the role played by federal-provincial power sharing arrangements, the heavy reliance on discretion in Canadian law, the role of the courts, and the impact of social norms. Working from a broad perspective that incorporates the insights of economics, law, political science, and philosophy, Green unpacks the features of Canadian policy making that determine the successes and failures of climate policies. His message is ultimately optimistic: *Picking Up the Slack* sheds light on how we can bring about meaningful movement towards a fair and positive future.

ANDREW GREEN is a professor and the Metcalf Chair in Environmental Law at the University of Toronto Faculty of Law.

Dewey Chang Photography

(in) UTP insights

UTP Insights is an innovative collection of brief books offering accessible introductions to the ideas that shape our world. Each volume in
the series focuses on a contemporary issue, offering a fresh perspective anchored in scholarship. Spanning a broad range of disciplines
in the social sciences and humanities, the books in the UTP Insights
series contribute to public discourse and debate and provide a valuable
resource for instructors and students.

For a list of the books published in this series, see page 297.

PICKING UP THE SLACK

Law, Institutions, and Canadian Climate Policy

Andrew Green

UNIVERSITY OF TORONTO PRESS
Toronto Buffalo London

© University of Toronto Press 2022
Toronto Buffalo London
utorontopress.com

ISBN 978-1-4875-4751-6 (cloth)   ISBN 978-1-4875-5621-1 (EPUB)
ISBN 978-1-4875-5011-0 (paper)   ISBN 978-1-4875-5331-9 (PDF)

Library and Archives Canada Cataloguing in Publication

Title: Picking up the slack : law, institutions, and Canadian climate
    policy / Andrew Green.
Names: Green, Andrew, 1964– author.
Series: UTP insights.
Description: Series statement: UTP insights | Includes
    bibliographical references and index.
Identifiers: Canadiana (print) 20220203237 | Canadiana (ebook) 20220203377 |
    ISBN 9781487547516 (cloth) | ISBN 9781487550110 (paper) |
    ISBN 9781487556211 (EPUB) | ISBN 9781487553319 (PDF)
Subjects: LCSH: Climatic changes – Law and legislation – Canada. |
    LCSH: Climate change mitigation – Government policy – Canada. |
    LCSH: Climatic changes – Political aspects – Canada. | LCSH: Climate change
    mitigation – Economic aspects – Canada.
Classification: LCC KE3619.G72 2022 | LCC KF3783.G72 2022 kfmod |
    DDC 344.7104/6342–dc23

We wish to acknowledge the land on which the University of Toronto Press
operates. This land is the traditional territory of the Wendat, the Anishnaabeg, the
Haudenosaunee, the Métis, and the Mississaugas of the Credit First Nation.

University of Toronto Press acknowledges the financial support of the Government of
Canada, the Canada Council for the Arts, and the Ontario Arts Council, an agency of
the Government of Ontario, for its publishing activities.

Canada Council   Conseil des Arts
for the Arts   du Canada

ONTARIO ARTS COUNCIL
CONSEIL DES ARTS DE L'ONTARIO
an Ontario government agency
un organisme du gouvernement de l'Ontario

Funded by the   Financé par le
Government   gouvernement
of Canada   du Canada

Canadä

Contents

Preface

When you are studying any matter or considering any philosophy, ask your-self only what are the facts and what is the truth that the facts bear out. Never let yourself be diverted either by what you wish to believe, or by what you think would have beneficent social effects if it were believed. But look only, and solely, at what are the facts. That is the intellectual thing I should wish to say.

The moral thing I should wish to say to them is very simple. I should say love is wise, hatred is foolish. In this world, which is getting more and more closely interconnected, we have to learn to tolerate each other, we have to learn to put up with the fact that some people say things that we don't like. We can only live together in that way. And if we are to live together and not die together, we must learn a kind of charity and a kind of tolerance which is absolutely vital to the continuation of human life on this planet.

Bertrand Russell (1959), on being asked what advice
he would give to future generations

I realized one summer afternoon that I live my life with a sense of unease. It is not that I am constantly fearful or depressed. In fact, on most days I spend much of my time living in the past – before I knew about climate change, about the costs I was imposing on others. I drive when I want, I enjoy eating burgers, I buy the new-est gadgets. I enjoy those days, basking in blissful ignorance.

But it is like sitting enjoying the hot sun and looking across a lake to see swelling thunderclouds. You hope the clouds will blow by but you have a feeling they won't. I know climate change is coming (and in fact is already hitting). Yet some politicians say it is too expensive to act, that it would make our lives harder and would be pointless. We cannot afford to act, at least until we have paid off our pandemic expenses. Others promise that we can have it all – reduced emissions, economic growth, Indigenous reconciliation – but do little that brings actual change. They seem to be talking past each other or, worse, complaining that those "on the other side" are mean-spirited or ignorant or self-serving. Change seems impossible, the way ahead stormy.

Many Canadians seemed stuck in this same uncomfortable position – enjoying the fruits of our growing economy while sensing that things have to change. We are frozen by a natural inclination to hold onto what has given us so much and a fear that change means losing it all. We need to find a way to move forward. It will not be by prodding people with dire warnings, as there have been plenty of those and they do not seem to be working, at least not for me. We need to think about why we fail to act, why we do not push for more from ourselves and our governments, why our governments do not actually seem to be acting in our best interests. We have to find a way as a country to take steps to avoid the harm we can see coming.

So, I started writing. While I was busily drafting arguments about how climate action would affect the economy and how we can think about the fairness of climate policy, the country seemed to be waking up, at least somewhat. Climate marches broke out across Canada; Greta Thunberg, a Swedish teenager, took over the news cycle; politicians started to fall over each other in proclaiming their desire to address the climate crisis. Parliament officially declared a climate emergency. And more plans and some policies started appearing – money for this, more promises of that.

But, as political scientist Kathryn Harrison notes, Canadian environmental law has come in waves.[1] Our progress on environmental issues has been propelled by public concern, but when that concern fades so does momentum. Even as climate concerns were hitting the front pages, it was not clear that this was a real wave. While Canada's economic picture was generally good, this was not the case everywhere in the nation; Alberta, in particular, was just emerging from a downturn in its fortunes. Moreover, many provinces then elected governments that were focused more on the economy than on climate. A looming wave seemed to stall.

And then came two shoals. First was a fight between two grasping oil giants – Saudi Arabia and Russia. When they failed in their bid to agree on how to divide their immense potential wealth from oil sales, Saudi Arabia decided to strike out at Russia, flooding the market with oil. The result was a crash in global oil prices. The Canadian oil and gas industry reeled and the stock markets began to tumble.

But the bigger shoal was partially hidden at the time of the oil price crash – the pandemic. It took some time for the world to realize the size of the COVID-19 crisis. When it did, countries shut down and so did the global economy. Stock markets fell at a rate not seen before. Governments in Canada and elsewhere sought to soften the impact, but it is hard to keep an economy afloat when you close down huge swaths of activity. And the drop in activity meant oil prices fell even further. In March 2020, in the early stages of the crisis, the Parliamentary Budget Officer foresaw a real GDP decline for Canada of 5.1% (the weakest since 1962) with unemployment rising to 15%, assuming the social distancing measures stayed in place for six months.[2] Things only got worse, with most countries starting to improve but then caught in a second and a third wave of infections and more despair, isolation, and death while waiting for a vaccine.

Did the pandemic end the wave of climate concern? People were understandably concerned about their jobs and opportunities for their children. However, the pandemic did not change the nature of what we have to do; it made some things harder but others potentially easier. First, some things changed but much remained the same. Economic hardship was felt across the country, but it hit some places and people more than others. As before the pandemic, we still need to think about how climate action will affect different types of industries and regions. We still need to consider how addressing climate change will affect different individuals – the less well off as opposed to the better off, rural versus urban

Canadians. And we still need to build institutions that reach fair, legitimate, and effective solutions.

Second, as we will see, we can learn much from the COVID-19 pandemic. There are many similarities between the pandemic and climate change. They both affect everyone in every country; they both require us to confront the economic effects of addressing a crisis; they both involve some initial denial followed by a growing recognition that action is required; they both create risks that are worse for some because of inequality; they both require changes in how we live our lives to address them.[3] These similarities can help us see more clearly the solutions that may work for climate change. They show us how government can react to a clear crisis, how individuals can pull together, and how industry can help.

We can also learn from the differences. At least in the beginning stages, governments did not have strong entrenched interests that opposed action on COVID-19; action on climate change, on the other hand, brings conflicts in economic interests front and centre. There has been (with some exceptions) wide-spread trust and faith in scientists and experts in the case of COVID-19; climate change has seen much more disagreement. COVID-19 brought with it grim pictures of its imminent threats; climate change has been tied to some examples of storms and fires but much of what drives action seems more remote. These differences give us the ability to think about the obstacles to acting on climate change.

Finally, the pandemic provided an opening. If we are going to spend vast amounts of government money to tackle the economic impacts of COVID-19, we may be able to use it to set up the conditions for addressing climate change.[4] The pandemic exposed the need to strengthen the social safety net and create new infrastructure to address health and other concerns and generated demand for support for ailing businesses. All provide the opportunity to move in a climate-friendly direction. Of course, how quickly this can be done and how we set about it are controversial questions. Those who have lost their jobs or are facing bankruptcy are understandably concerned if they believe their losses are ignored or overlooked. We need to build the trust and agreement necessary to bring everyone along.

Even though we ran into a merciless pandemic, then, the central concerns remained how we can address climate change in an effective, fair, and timely way.

"The More You Explain It, the More I Don't Understand It" (Mark Twain)

I set out my main arguments in chapter 1, but first note that this book sits in a somewhat uncomfortable spot. I have tried to strike a balance throughout between technical depth and accessibility. To find the balance, I have laid out the basic facts and arguments but left the support to the endnotes for those who want to get into the details. My aim was to build off recent findings from economists, political scientists, lawyers, and other experts. There is a tremendous amount of work on climate policy in Canada that has not been drawn together. I have spent my academic life working on environmental and related problems, including climate change, from a legal and economic perspective. My work has attempted to bring together various disciplines, including law, economics, philosophy, and political science, to study how we can bring about change, how we can use law to foster action, how institutions (including the law, norms, and markets) work. If I wanted to help with climate change, the best way was to bring what I have to the table by laying out what we know and what it could mean.

But I also wanted this work to be set out in a readable manner that provides the contours of a solution. I wanted the book to be short enough that anyone with an interest would be willing to pick it up. Everyone is so busy that it is hard to invest huge amounts of time in any one subject, even one as important as climate change. There are other excellent books out there that cover the science of climate change or the details of different policy choices or the ethical demands of action or the role of international institutions. I tried to draw on all this prior work to build an accessible story and to show a path forward for Canada using what I know about how law and Canadian institutions work.

I've also tried to find common ground. This book does not hit hard on political themes. I worry that we fail to find workable, effective solutions because we get lost in the rhetoric of blame and the anger that often goes along with it. We shut down, stop listening to each other. I wanted this book to lay out the facts as I see them and suggest a way forward. I believe we need strong action on climate change, but I also believe that we will only get there by listening to each other and thinking about how we can find a shared way forward. We seem at times to have lost faith – in our government but also in each other. We need to find trust and empathy to get out of the mess we are in.

The puzzle lies in why it is that we can know about the impending dangers of climate change and do little about it. Part of the answer, I believe, is to look squarely at where we have come from – both the good and the bad aspects of what we have built in Canada. This is the only way we will be able to find the footing to work towards a fair, positive future for all of us.

Many Views, One Cathedral

One of my favourite article titles is "Property Rules, Liability Rules and Inalienability: One View of the Cathedral" by Guido Calabresi and Douglas Melamed, two famous originators of the field of law and economics.[5] The phrase "One view of the cathedral" is said to refer to Monet's series of paintings of Rouen Cathedral showing different details and angles. Calabresi and Melamed were saying: here is one way to look at this issue of private law but there may be others, each with something different to say about the actual cathedral. In the article, they point out that while frameworks allow you to see relationships you might otherwise miss,

> framework or model building has two shortcomings. The first is that models can be mistaken for the total view of phenomena, like legal relationships, which are too complex to be painted in any one picture. The second is that models generate boxes into which one then feels compelled to force situations which do not truly fit.[6]

This is a good way to think about academic theories generally, but work on climate change in particular. Every academic discipline seems to have its own perspective on what needs to be done to address climate change and why Canada hasn't taken sufficient action. Economists often view climate change as involving the choices of rational actors and try to find a solution that maximizes some notion of the welfare of the country. Philosophers may disregard this outcome-oriented approach to try to reveal other reasons to support certain actions that are based more on, for example, the distribution of the benefits of society. Political scientists, and often lawyers, may sketch how the institutions frame decisions and give effect to certain values over others. The views heavily overlap, but they emphasize certain features and shade over or omit others.

One of the tasks of this book is to try to use these different views to get a clearer picture of the carbon challenge facing Canada and what we can do about it. We will need to see the problem from a number of these perspectives, but we can't possibly delve into each one: we would never have the time or ability to understand the problem as a whole. Some issues, such as the role of international institutions and global justice, will be mentioned, but the book will focus mainly on the Canadian context and institutions, which is enough of a task. The best I can hope for in this short book is that we get a rough sense of enough current perspectives that a better appreciation of the cathedral emerges.

One important aspect of the Canadian context generally but also with respect to climate change is the rights of Indigenous peoples and the history of their relationship with Canadian governments. It underlies discussions of environmental justice; of the impacts of different policy choices; of governance institutions, including self-governance of Indigenous peoples; of trust; of the role of law and the Constitution; and more. In effect, it touches on almost every core element of this book. I have tried to weave discussions of Indigenous rights and perspectives throughout these pages but, given the scope and complexities of the overlapping issues of Indigenous rights and climate policy, the discussion is unfortunately incomplete. The topic could be, and deserves to be, the subject of another entire book.

As mentioned above, Canadian environmental law has developed in "waves," with public concern pushing forward change only for the wave to crash as the public becomes concerned about something else, usually the economy. We may have felt the rising of a "climate" wave, but one worry is that it is crashing like all the others on the shoals of a recession and the need for growth. Even before COVID-19, we were seeing some politicians state that while climate action is important, we just cannot afford to act right now as we face an uncertain economic future and an unwilling trading partner to the south. The downturn caused by the pandemic and more recent events made those calls louder, and some of them have been answered with policies that weaken climate action.

A related worry is that the wave of public concern will continue but will roll over an immovable and unyielding shore. In the past, politicians have made bold statements about the need for action on the environment and even taken action that looks substantial. In reality, though, this action either never happens or is undercut by exceptions and exemptions. The wave rolls on by, people forget, and things continue as before.

You cannot make good policy this way. This book aims to give a sense of what is actually happening on climate change in Canada. People make claims about climate action – that it will kill the economy, that it will hurt the poor, that we have taken strong action, that we have not taken any action. We need to see what the experts have discovered about these claims. This will give us the basis to talk about our possible policy options. What can we do that makes sense given what we know about climate change and the steps we can take to reduce emissions?

I have a tall order ahead of me – to follow the advice of Bertrand Russell I set out at the beginning of this preface. I want to start from the facts as they are, to be open to where they take me. And I want to try to empathize with all the variety of Canadians who look to the future with fear and with hope – both those who want strong, immediate action on climate change and those who dread it. It is likely too much, but I feel I need to try.

Acknowledgments

Many people have been so kind and generous in helping with this book – reading drafts, giving encouragement, discussing ideas. Two deserve particular notice. My brother David read multiple versions of different chapters and critiqued as only an older brother can. More importantly, though, he has been a constant source of encouragement and support, not only for this book but for my whole life. I cannot begin to pay him back. The other is Michael Trebilcock. He was immediately enthusiastic when I started this project, and he suffered through many early drafts as I sought to find a path forward. As he has been throughout my career, he has been a model of what a generous colleague should be. I am very fortunate to have others who have read drafts and were willing to discuss the ideas in the book, including Albert Yoon and Ben Alarie, whom I have been very fortunate to count as friends and collaborators over the years.

I also wish to thank the thoughtfulness, passion, and hard work of the students who have helped with the book, including Yara Wilcox, Justin Mayne, Declan Walker, Jaclyn Medeiros, Cristin Hunt, and Adam Davis. I am grateful as well to my students, on whom I inflicted some draft chapters and who were willing to provide helpful comments. This work was supported by funds from the Social Science and Humanities Research Council, the University of Toronto Faculty of Law, and the Metcalf Chair in Environmental Law.

At U of T Press I am immensely grateful to Dan Quinlan, who was extremely helpful, carefully reading the book and providing wise advice, and to Terry Teskey for her thorough editing. I would also like to thank two anonymous reviewers for their thoughtful and constructive comments on the manuscript.

Finally, of course, I wouldn't be able to do any of this without my long-suffering wife and daughters. They lived with the complaints, the worries, and the doubts and were consistently enthusiastic (or at least were willing to fake enthusiasm). They were a constant source of support and ideas. I cannot begin to express my appreciation and love.

PICKING UP THE SLACK

Law, Institutions, and Canadian Climate Policy

chapter one

Of Fear and Loathing in Canadian Climate Policy

The environment sits in an odd place in Canada. Canadians identify with the outdoors and nature yet allow wild areas to disappear. Canada has become rich from natural resources yet does not
provide for their long-term sustainability. As Canadians, we see
ourselves as nurturers of the forests, waters, and animals, yet we
act like simple extractors. Like most countries Canada does some
good things and some bad, with the result that overall it ends up
mediocre on its environmental record. The 2020 Yale Environmental Performance Index (EPI), for example, ranks us 20th out of 180
countries:[1] so, good but trailing countries like the United Kingdom,
France, Norway, and Sweden who might be thought of as comparable. Canada does well on overall air quality (8th) and reduction
in sulphur dioxide emissions (1st), poorly on species habitat protection (104th), tree cover loss (101st), and fisheries (89th).

Climate change has not been Canada's strong suit. The Yale EPI
ranks Canada at 37th overall in addressing greenhouse gas (GHG)
emissions, slightly better on limiting the growth of carbon dioxide
emissions (33rd), but very poorly on GHG emissions per capita
(168th), growth of nitrous oxide (a much more potent GHG than
carbon dioxide, arising from agricultural and industrial processes
and from fossil fuel use) (92nd), and in the intensity with which we
emit GHGs to produce things (77th). Canadian governments have
drawn up targets to reduce emissions but have consistently missed
them. Instead they have set new targets farther out and in general
have been unable to limit release of GHGs.

In some ways the difficulties Canadians face in agreeing on and meeting targets are not surprising. Canada has grown wealthy from both the production and use of fossil fuels. It has tremendous natural resources. Oil, gas, and coal have driven prosperity and jobs in many different parts of the country. Canadian forestry and agricultural sectors have relied on fossil fuels and have generated their own GHGs. Cities and supply chains have been built on fossil-fuel-based transportation. In some parts of the country, fossil fuels generate much of the electricity.

At the same time, the science is clear. We need to reduce emissions if we are going to avoid disastrous consequences. Increased warming could mean more extreme storms, heat waves and droughts, fires, displaced people, negative health impacts, food insecurity, economic disruption, and more.[2] Canada faces harm to coastal and northern communities and to Indigenous communities generally, more fires, harms to human health, changes to habitat for animals, disruption of shipping and hydropower, risks to forest productivity, increases in pests, and on and on.[3] The impacts are environmental, but Canada also faces "increasing economic costs" from effects both in Canada and abroad.[4] The costs of not acting are high.

Change is happening. More electricity is coming from renewable, non-carbon sources all the time. The federal Liberal government announced new plans in March 2022 claiming finally to have a package of policies that will enable us to meet our international commitments. The provinces are taking measures to address GHG emissions, some stronger than others. Reports seem to be coming almost daily of companies making new pledges to reduce emissions or invest in cleaner technology.[5]

Yet it is not enough. Under the Paris Agreement, Canada's most recent international commitment to address climate change, the federal government committed initially to reduce GHG emissions to 30% below 2005 levels by 2030. Unfortunately, the UN estimated that even if all countries met their 2030 targets it would not be sufficient to meet the agreement's initial goal of keeping temperature increases below 2 degrees C, with a stretch goal of below 1.5 degrees C.[6] And even more unfortunately, most countries are not on track to meet these commitments. The federal Liberal government strengthened its target in 2021 to reducing emissions by 40–45% below 2005 levels by 2030, which it claims it can meet with its newest proposed policies. Moreover, it enacted legislation that sets a 2050 goal of net-zero emissions – so we remove as much in GHGs as we add.[7] However, the nearer-term 2030 commitments are less than the US commitment of cutting emissions by half and EU and UK commitments of 55 and 68%, respectively.[8]

Further, the government's plans are just that – plans. It may be being pushed by public pressure for climate action, but there have been many plans before that have not come to fruition. In fact Canada's emissions *rose* by 1% between 2015 (when the Paris Agreement was signed) and 2019, the only G7 country to have increased emissions.[9] Moreover, some have expressed grave doubts whether the federal and provincial plans will be able to bring us anywhere near the promised reductions.[10]

The core of the problem arises not from climate change itself, although that shows the problem in its starkest light. The story of Canadian law and policy relating to the environment and natural resources has involved periods of at best indifference and at worst devastation, punctuated by brief bursts of new laws and government agencies focused on environmental protection. Political scientist Kathryn Harrison pointed out back in the 1990s that Canadian environmental law has come in waves.[11] When the public became concerned about chemicals, Canada got its first modern environmental laws. Another wave came in the early 1980s and then again in the 1990s. Each wave rose on greater public concern and brought with it more enhanced environmental legislation, but each crashed when faced with economic downturn. The public and politicians turned from fixing the environment to supporting workers and getting industry going again.

This pattern has continued since that time. Canadians generally appear to be concerned about climate change and seem to support some action. However, even pre-pandemic there was some doubt about how much governments would do. The Canadian economy generally was doing well but some, such as Alberta in particular, were recovering from prior downturns. Conservative governments in a number of provinces seemed to be against strong climate action. The oil price crash in early 2019, due to both a Saudi Arabia-Russia oil dispute and then the pandemic, knocked the Canadian economy back on its heels. As the pandemic raged, however, the federal government and some provinces promised even stronger action.

Should we believe that the pattern of avoiding effective action can be changed? The economic consequences of the pandemic and the oil price drop have been devastating: people, entire industries, and provinces were hurt. Moreover, the federal government seems at times to neglect working with the provinces in coming up with the direction of its plan. The hope is that policies will drive change while growing the "green" economy. However, the key question remains: how can we tackle a crisis like climate change that we see coming but that involves such conflicting and complex interests?

The Common(s) Problem

We will work from the assumption that we need strong action on climate change and we need to start now. That of course says nothing about what that action should look like. Chapter 2 provides an overview of the science, along with what Canada has (and more to the point has not) done so far. Canadian governments have said emissions must be reduced, promised they would be, but have so far not been able to fulfil those promises. Why? As we will see, it is not for lack of tools to bring about emissions reductions. In his most recent book, economist Mark Jaccard persuasively argues that between flexible regulations and carbon pricing, we have all we need to hit the targets if we want.[12] Yet Canadian governments do not use those tools in an effective manner, and undermine them with exceptions and delays when they do use them.

Nor do we lack creative, thoughtful ideas about the limitations of the current system and what they mean for climate action. Naomi Klein, for example, has for some time been challenging the market system and decrying its implications for the climate. She urges us to think about what Canada's current reliance on narrow, economic-focused thinking means for people. You may not agree with her solutions, but she does expand the universe of ideas about where the problem lies.

This book builds on work by economists, political scientists, and others to think about what it will take to bring about action in the current Canadian context. Legal scholar Michael Trebilcock points to three elements that either help or hinder policy: ideas, interests, and institutions. "Ideas" are about what we should be doing, what our goals or ways of thinking about a problem should be. "Interests" refers to all the parties who have skin in the game – who care about the outcome and try (or do not try) to steer policy to foster their own interests. "Institutions," such as how governments make decisions or how the country is set up, influence whose interests and ideas prevail.

My main message is about Canadian institutions. The basic argument is that Canada has failed to act because of wrong or at least overly narrow ideas about what to do, because of the conscious creation and maintenance of strong interests that do not see climate action as beneficial, and because of weak institutions that are unwilling or unable to act in the face of these interests. Thinkers as seemingly different as Naomi Klein and Nobel-prize-winning economist Amartya Sen have written about how we take too narrow a view of what our goals should be, focusing too much on economic growth as the metric of our success. Economists and political scientists have pointed to the strong economic interests of industries that produce or rely on carbon. These industries have had considerable sway in policy development around the world, particularly in Canada. And as individuals we are comfortable with the lives we have become accustomed to, based on the bounties of carbon.

Institutions, however, are at the core of Canada's particular difficulty in addressing climate change. Of necessity, institutions have to be somewhat place-based – we have to be culturally and historically minded about where our institutions come from, what they are good at, and where they let us down. This focus allows us both to see where the problems lie and to craft a solution. My argument tries to look carefully at the institutions Canada has – like how laws are made, how power is divided up in Canada, what types of policy solutions have been adopted – to see why Canada has arrived at this precarious spot.

These ideas, interests, and institutions overlap and feed off each other, but together they have hindered and will continue to hinder climate action. I set out the basic nature of the problem that climate change poses in chapter 2. I call this part the "common(s)" problem for two reasons. First, as is well known, climate change is at its core a commons problem – individuals, countries, and provinces bear the costs of reducing emissions but share the benefits with everyone else. The result is everyone has the incentive to let others bear the costs of action while enjoying the benefits. If everyone acts this way, nothing gets done.

Second, however, the problem at the core of climate change is a common problem for Canadian environmental and natural resource law, just more extreme. After Europeans landed on Canada's shores, fish, fur, and forests formed the basis for trade with other countries and for economic growth. These "staples" were soon augmented with wheat, minerals, coal, and eventually oil and gas. The importance of these different industries to jobs and growth shaped Canadian laws around natural resources and the environment. Canada has struggled to find institutions that allow both economic growth and a healthy relationship with the environment and with Indigenous peoples. As we will see, climate change policies and laws are no different, although the need for immediate, credible action adds an element of urgency.

Addressing climate change, then, is difficult for Canada both because of the commons problem that afflicts individuals, provinces, and countries and because of Canada's particular place in the world, such as its history of heavy reliance on economic returns from natural resources, in particular fossil fuels. To get a grasp on what needs to be done, I break our discussion of climate change into two broad parts: why Canadian institutions are failing us and what can be done about it.

Why Canadian Institutions Are Failing:
Discretion, Diffusion, and Deference

As mentioned, Canadian climate policies stem in part from overly narrow ideas about appropriate policy goals and from extremely strong interests that have benefited from laws and policies that foster economic growth. The second part of the book discusses how Canadian institutions have both fostered these ideas and interests and failed to create pathways to resolving conflicts with other values such as environmental protection. It focuses on three related problems: the vast amount of *discretion* that lies at the heart of climate change laws and policies; the *diffusion* of responsibility for climate action; and the *deference* to government decisions on climate by the courts and by the public.

Discretion. Chapter 3 starts with the basic idea that Canada is not missing its targets because of a lack of tools to meet them. Economists, political scientists, engineers, and others have fleshed out the relative cost and effectiveness of carbon pricing versus regulations or subsidies. We have a good idea of what can work. The federal and provincial governments have set up their environmental laws to allow them to take strong action using any and all of these tools.

Yet Canadian policy has tended to rely more heavily on less effective subsidies and information policies. Even when governments use pricing or regulation, which are seen as more effective, they generally water them down: the requirements are made weaker and breaks are given to a wide range of industries. If we use even relatively conservative estimates of the potential costs of climate action, Canadian policy seems to be based on hope – hope that the worst effects of climate change will not happen.

The source of the problem lies in how we make these policies. Canadian laws give those who make climate policy very broad discretion – to decide what the carbon price should be or set the emissions levels from particular industries or approve new pipelines. This is for good reason: discretion provides space for policy decisions to be based on the latest science and to adapt as new information arises. The difficulty, though, with giving someone all this discretion is that they may use it in less than publicly spirited ways. And we need to figure out when that happens.

Why might Canadian climate laws not be as tough as it seems climate science demands? Chapter 4 starts with the toughest obstacle – the claim that Canada cannot take stronger climate action without destroying its economic future. According to most studies, however, while climate action will have a negative impact on the economy, economic growth will continue, just a little less than would have happened otherwise. The difficulty is that some industries will be hurt more than others. The key will be how "emissions-intensive" various industries are (how much they emit in producing whatever it is they produce) as well as how "trade exposed" they are (how much these industries compete with those in other countries). Other countries may not be taking significant action on climate change, and so if Canada imposes costs on its industries, they may be at a competitive disadvantage compared to industries in those other countries. As a result, Canadian industries may produce less or perhaps even move overseas.

This exposure to competition with other countries is one of the trickiest issues in discussions of climate change policy, as it raises considerable fear, both legitimate and not. Canada does have a few potentially important industries that are both very emissions intensive and "trade exposed," but that is not true for all industries. Canadian climate policies in part are weak because policymakers are attuned to the importance of these costs to industry, in particular the oil and gas industry. We may want to help some industries out – by exempting them from certain climate policies or giving them time to adjust or paying them to reduce emissions. Yet rather than focus support on those industries that really need it, Canadian governments have tended to try to help all industries. Moreover, in many cases the requirements have been significantly weakened for industries, such as oil and gas, where they would have the greatest impact. One explanation goes back to our institutional story – that Canadian policy-makers have a tremendous amount of discretion about what policies to impose, and they have used it in a misguided attempt to make "feasible" policies.

Diffusion. The second concern about the ability of Canadian institutions to deal with climate change lies with the diffusion of responsibility across different levels of government. While such diffusion can be a source of strength, it has not worked out that way overall in reducing emissions. To understand the scope of this problem, chapter 5 takes our work on differences across industries and extends it to provinces. Strong action on climate change will be much harder on resource-based provinces or provinces that rely more heavily on fossil fuels like coal. Alberta is the most obvious example. Albertans are naturally concerned that reducing national emissions will mean that they will bear the brunt of a policy from which all of Canada benefits. This concern about the fairness of climate policy for different regions has been the source of considerable anger and debate in Canada.

There are different ways to parcel out emissions reductions across Canada, such as reducing where the costs are lowest, requiring the same emissions reductions per capita in each province, or having each province reduce emissions by the same percentage. Whatever you think of these different principles, Canada has no real coherent plan based on any of them. The federal government set a national target and is trying to impose some minimum standards on the provinces that are insufficient to reach the target on their own (although the federal government's most recent plan is stronger). The provinces are left to their own devices to choose whatever targets they want – some of which seem to align closely with the national target, others not so much.

The lack of a fixed concept of how much each province should reduce brings with it flexibility but no assurance Canada will meet climate targets in any fair manner. There are other options. The federal government could try to legally apportion emissions reductions by province. As we will see in chapter 9, such a plan may not be permitted under our Constitution. A plan could, on the other hand, build on Canada's history as a risk-sharing enterprise. At various times in our history regions that were doing well have helped those that were struggling because of different crises in their resource or manufacturing sectors. Climate change is just another one of these crises, yet it has not really been seen as such.

As a nation, Canada has encouraged investment in the fossil fuel industry in order to create jobs and to increase tax or royalty revenue to fund government programs. It seems only fair that when we try to reduce emissions from this sector, everyone helps pay for it. Canada is fortunate as it is set up to deal with, and has experience in, just such situations.

Deference. Canadian climate change polices are, then, built on laws that delegate broad discretionary powers and a constitutional structure that divides responsibility across different levels of government. Discretion and diffusion are not necessarily bad – they can bring flexibility, experimentation, and expertise that can lead to better policies. However, chapter 6 sets out how Canada lacks the controls necessary to ensure this discretion and diffusion of responsibility is not misused.

We can gain some confidence in government decisions if they are based on *expertise* (on science for example) or made in a *politically accountable* fashion or made in line with some *legal* principles or rights that we feel are necessary, such as the freedom from discrimination in the Charter. It turns out all three are needed for the system to work; their strengths and weaknesses offset each other. Political accountability, for example, may bring a connection to the values of Canadians but lacks the expertise to make complex decisions or the objectivity that can come from the courts.

Unfortunately, behind Canada's weak climate policies lies an over-reliance on political accountability. Cabinet or ministers make many of the decisions that are central to strong climate action, but there are few effective checks to make sure those decisions are rational and in the long-term interests of Canadians. The discretion given to these officials allows them to override expertise. Moreover, the courts have been extremely deferential to these government decisions, especially in areas such as climate change that rest on science. The result has been promises but insufficient action.

Finding Balance

Discretion, diffusion, and deference create space for policy to move away from the public interest – what some have termed "slack."[13]

The title of this book points to the way forward – thinking about how to reduce that slack. We have heard plenty of "solutions" to help Canada address climate change – there should be more mandatory rules; the federal government should take control; the decisions should be based on expertise; citizens need to hold leaders politically accountable; there should be more public participation; governments need to lead and the public will follow; and on and on. Each of these contains elements from which a stable solution can be constructed, but part of the problem has been a failure to consider the whole picture. Chapters 7 to 11 tie these pieces into the issue of institutional failure and the path forward. The solution lies in focusing on people, strengthening the national community, cultivating cooperation, fostering trust, and setting a solid foundation through norms.

Focusing on People. One argument made against strong climate action is that it is unfair because it will disproportionately hurt the working poor/middle class or workers in carbon-based industries. Any policies to reduce emissions are going to impose a cost on emissions, either directly (through a carbon price) or indirectly (by requiring industries or individuals to take costly action). The price of food, gas, heating, everyday things will go up. Jobs will be lost in some industries.

Chapter 7 looks at the issue of whether the costs of climate action will be unfairly visited on some, especially vulnerable, individuals. Canadian governments need to pay attention to these costs both out of fairness to all Canadians and, pragmatically, to address the need to reduce opposition to climate action. Workers in industries that will be harmed by strong climate action, for example, naturally fear the effects of the transition to a "net zero" economy. To find the political will to support strong action, these fears need to be addressed.

It turns out Canadian climate programs do not tend to impose added costs on the less well-off but there are real risks for workers in carbon-based industries. To get people to see that climate action is in their interests, we need to think of what Canadians really want – money or jobs for sure, but more. Nobel-prize-winning economist and philosopher Amartya Sen speaks of the need to ensure people have the freedom to live the lives they have reason to value. For Sen, this means ensuring they have economic freedom but also other things such as political freedom, education, and security from risks to health or employment.

On this view, the issue is not just one of making sure that the less well-off are not hurt by climate policies but more of thinking broadly about making them better off. For workers in carbon-based industries this goes beyond the usual "re-training" programs. Steps need to be taken that overcome people's fear not only of not being able to find jobs in industries that do not yet exist but also of losing their identities and communities. Failure to do so means an understandable mistrust of and sense of unfairness in climate action. Canadian governments have failed to make the case that everyone has an interest in making the transition. There is a positive story to tell – that if we look at what people want more broadly, the transition can be in their long-run interests and Canada has the institutional framework to share the more immediate risks.

Strengthening the National Community. In chapter 8 we look at a related argument about fairness: that Canada cannot take strong climate action as it will mean governments will lose so much revenue they will no longer be able to afford important social programs like universal health care or public education or social assistance. Granted, the fossil fuel industry, for example, is important to government revenue – but not overwhelmingly so (although there are some exceptions, such as Alberta). Moreover, Canadian governments pay out a fair amount in subsidies to the fossil fuel industry, which does not completely offset the revenue governments receive from them but does put a big dent in it.

Canada has developed ways to ensure all provincial governments can fund public services for all Canadians. The federal equalization program is focused on each province having essentially equivalent resources available to fund government services, and the stabilization program aims to ensure that when a province is hurt by a downturn, federal resources support it temporarily. Federal programs such as the Canada Health Transfer directly support provinces in providing some services for all Canadians. Again, this is a system based on sharing risks. These programs have been under attack as the source of concern with climate programs – that climate action will hurt provinces that pay into the system, and these programs do not pay out enough to make a difference. But these risk-sharing programs are the source of the solution, not the problem. They need to be adjusted to share the risks from climate action – this is essential to persuading everyone that action is both the right thing to do and is in their interests.

Cultivating Cooperation. Chapter 9 asks whether the federal government should just step up and tell the provinces what to do. If the provinces are all waiting for some other province to reduce emissions, shouldn't the federal government just intervene and set the course? That is, after all, a clear way to solve disputes between parties; someone else imposes a decision.

Canada was not set up that way, however. The Constitution does not assign power over "the environment" to either the federal or provincial governments. The Supreme Court of Canada has essentially said both have the power but need to use it in a way that respects the other. Just like Canadian policies aimed at risk-sharing for government finances, this structure is the solution, not the problem. Action is needed at all levels. Provinces have local knowledge and legitimacy; the federal government can step back from local interests and make decisions that may be less swayed by industries that are important to a single province. Both are important and necessary for climate action, and both are necessary to promote trust in a solution.

Fortunately, the Supreme Court has interpreted the Constitution so as to promote cooperation between the federal government and the provinces. Cooperation is key to effective Canadian policies in areas that seem both national and provincial. Public health care was set up through an agreement in which the federal government transferred money to provinces as long as they met certain conditions. Canada enacted its own "new deal"–type policies such as unemployment insurance in part through getting agreement on a constitutional amendment. Canada has moved forward through compromise and cooperation. These also set the path for our transition. They allow for coordinated action between the federal and provincial governments and are necessary elements for reconciliation and agreement with Indigenous peoples.

Fostering Trust. All this action requires trust – trust in governments and in each other to do the right thing. Chapter 6 discussed a key concern that the system leans far too heavily on political accountability – and in particular on political accountability through Cabinet. The role of expertise and the courts as well as other forms of political accountability such as public participation and open consultation with Indigenous peoples has been neglected.

Chapter 10 discusses the search for balance. The role of expertise in the decision-making process should be enhanced. Independent bodies that have expertise and a broad range of experience must play a greater role in climate decisions at both the federal and provincial levels. The ultimate answer is unlikely to fall to these expert bodies, and should not, given the value trade-offs involved. But having an independent body take on the problem for the government to react to can set boundaries on possible responses.

In addition, Canadian courts must require more by way of reasons for decisions from the experts, but also from the politically accountable actors making the final decisions. Reason-giving can provide the basis for more effective accountability. Canadian legislation can also provide broader, specific goals for climate policy. Such goals provide the courts and the public a foothold to review climate policies. The idea is not that the courts should force governments to come up with a particular policy – since judges have neither the expertise nor legitimacy to make such choices. It would, however, mean that the government has to specifically account for what it decides.

There have been some tentative steps in this direction. Many environmental laws already require reasons from decision-makers, and some provincial laws and the federal *Canadian Net Zero Emissions Accountability Act* set targets in the legislation. The courts have in recent years placed a greater emphasis on reason-giving. But more is needed, as the current attempts at accountability measures reveal considerable gaps. Transparency is an important tool to reduce short-term or biased decisions.

Setting the Foundation. Finally, we come back to individuals again. Chapter 7 looked at policies to help people see addressing climate change as in their interests by tying climate action to their broader concerns, such as about jobs, inequality, and health care. Chapter 11 considers individuals' choices around climate change, both in reducing their personal emissions and supporting broader political choices to address the issue.

Central to changing behaviour and building the political will for climate policies are social norms – what people feel is right or appropriate. Norms can change as the result of a crisis, such as experiencing the extreme effects of climate change. This change is happening to some extent but not quickly enough. Policies and institutions can, however, make a difference. Information and education most obviously help to create a common ground for discussion of policies.

More broadly, policies such as carbon pricing or regulations not only change behaviour directly but can have an impact on norms. They provide a signal of what others in society think is important and can give confidence that others are willing to act. In addition, changes in individuals' choices about their cars or home heating or food can have a bigger impact: they make climate-friendly choices the norm. Such a norm can influence political choices – how the public holds elected officials responsible for making and meeting commitments on climate change. From a small trickle of individual changes can come a steady, consistent stream of political choices.

Finally, how laws are made matters. A common view of a path forward can be built through involving people in policy decisions. The processes can allow Canadians to debate and deliberate on difficult issues and forge a common set of values. It can provide opportunities for Indigenous voices. Currently Canadian policy processes often give only token opportunities for public involvement and little or no opportunity for deliberation.

The root of climate inaction is frequently argued to be lack of political will. At least part of the solution is to think consciously about how Canadian institutions and policies affect that will. We need to come to some common Canadian understanding of how the trade-offs necessary in the coming years should be made.

Changing the Cycle

There is the story: Canada has weak climate policies because it has weak institutions, relying too heavily on discretion with too few viable checks. Strong concentrated interests are able to influence policy decisions in such an environment, particularly where the main ideas behind policies focus on economic growth. Their influence not only can change individual policies but can shape laws and institutions. Some steps have been taken to address this concern, but not enough. Canadian governments have created forms of public accountability without setting up the foundation of support or the sense that change is in everyone's interest. They have tried to "strengthen" laws and create "legally binding" requirements without giving courts or other institutions an effective way to ensure these laws and requirements are met. They have created a greater role for experts but still allow considerable scope for their ideas and advice to be ignored or downplayed.

Individual choices can make a difference to emissions. Moreover, we seem to hear daily about how financial markets are putting pressure on companies to change. Markets will be central to driving innovation on emissions reductions and new forms of energy. Financial markets and other countries seem to be constantly pushing for change, limiting the space for unlimited emissions by companies. Yet individual action, markets, and even the influence of other countries are necessary but not sufficient. They will not be enough without the support of strong institutions. Action may not happen at all, but if it does, it will likely not be fast enough. Moreover, a transition to a net-zero emissions economy, if it happens, can happen in many different ways – a fair distribution of the costs and benefits of change or an even more unequal society. The pandemic is a case in point – governments moved to reduce the risks, but the path and outcome have not always been fair to all people. The same concern holds for climate change.

The solution lies in all three areas: ideas, interests and institutions. We need to broaden our *idea* of the goals of Canadian society – to focus on ensuring Canadians can share in lives that they have reason to want to live. We also need to ensure that the transition does not work against the *interests* of Canadians – to reduce the costs of provinces, industries, and individuals that will be hurt by the transition and increase opportunities for Canadians to benefit from the transition economically. Perhaps most importantly, we need to strengthen Canadian *institutions*, building on the strong base that already exists in sharing risks, in cooperation, and in trust.

These changes are both a lot and not enough. But they are a start. They are not an "all or nothing" solution. Elements are already in place in the basic Canadian structure, and more have recently been added. Momentum can build from greater accountability, an instinct for fairness, a focus on reconciliation.

chapter two

Cows, Cod, and Coal: The Roots of Canada's Climate Dilemma

This is not a book about the science of climate change. For one thing, I am not qualified to write about the finer points of cloud particles or ocean acidification. In any event, for me that is not really where the key debates lie when it comes to figuring out what Canada should be doing. We know the basic outlines of the problem and the risks. The truly pressing question is what to do given that we know.

However, we do need to be on the same page about what the science is telling us. The good news is that the existence of climate change no longer appears to be as much of an ideological issue – politicians on all sides claim to believe that climate change is happening and that it is being driven by humans. As others have set out fulsome accounts, I will not go into the history of climate science here.[1] It should suffice to say that since the 1990s, there has been a solid base of evidence of potential risks from climate change and that the evidence has only gotten stronger.

We keep hitting various records, but one that sticks out is that 2020 was tied with 2016 as the "warmest year on record" and "the last seven years have been the warmest seven years on record."[2] The warming of the climate since the Industrial Revolution has been "unequivocal," with global mean surface temperature increase estimated to be 0.85 degrees C between 1880 and 2012.[3] Canada is experiencing warming at "about double" the global average increase, in part because a loss of snow and sea ice is reducing the reflection of the surface and so increasing the absorption of solar radiation.[4] The effects are more dramatic in the Arctic, at about three times the global rate.[5]

The UN Intergovernmental Panel on Climate Change (IPCC) states we are in an "era of committed climate change."[6] While many different factors may be driving this warming, human activities are "extremely likely" to be the main cause.[7] We emit greenhouse gases from much of what we do – driving our cars, heating our homes, or making our food. Carbon dioxide is the most famous of these gases, but there are others that pound for pound are worse, such as methane. Some change in our climate is inevitable even if we drastically reduce GHG emissions. The warming of the planet from human activities "will persist for centuries to millennia."[8]

The key concern is future emissions, although we will talk a little about taking carbon out of the atmosphere. If GHG emissions growth continues as it has, average global temperatures are expected to go up by almost 4 degrees C by the end of the century.[9] Even if countries meet their emissions reduction commitments under the Paris Agreement, we would be on track for a 3 degree C increase by the end of the century, well above the Agreement's 2 degree C target, let alone the more recent resolve to keep increases below 1.5 degrees C.[10]

And emissions are going up. Figure 2.1 shows that emissions overall rose from 2005 and 2018. By the end of the period, the main emitters were China, the United States, and the European Union, with India increasing its share significantly over the period. The rising emissions were partly due to a strong global economy. By 2018, global energy consumption had almost doubled since 2010, with China, the United States, and India making up about 70% of the increase in energy demand. Connected with that increase was an increase in emissions. In 2018, for example, coal was the biggest driver of GHG emissions from energy use, accounting for about 30% of global emissions, and coal-fired electricity generation increased in both China and India by about 5%.[11]

We can see the connection between emissions and economic activity in struggling with the pandemic.[12] In February 2020, China's emissions dropped by about 25% as it shut down its factories to try to stem the spread of COVID-19. The International Energy

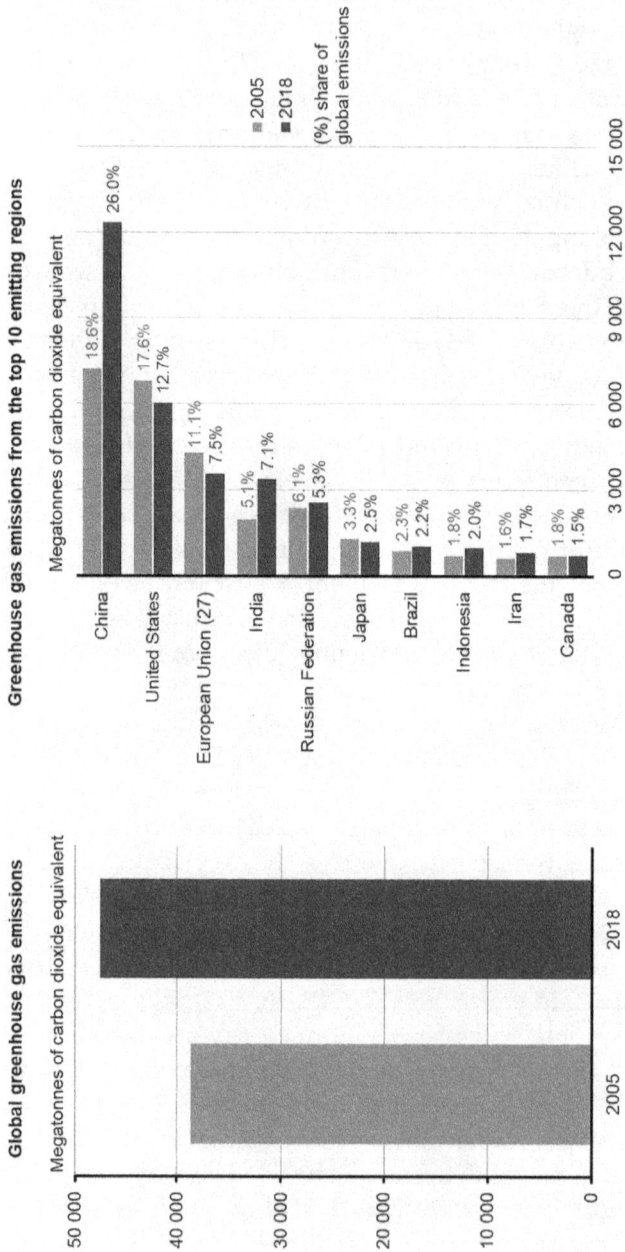

Global greenhouse gas emissions

Megatonnes of carbon dioxide equivalent

Greenhouse gas emissions from the top 10 emitting regions

Megatonnes of carbon dioxide equivalent

China — 18.6%, 26.0%
United States — 17.6%, 12.7%
European Union (27) — 11.1%, 7.5%
India — 5.1%, 7.1%
Russian Federation — 6.1%, 5.3%
Japan — 3.3%, 2.5%
Brazil — 2.3%, 2.2%
Indonesia — 1.8%, 2.0%
Iran — 1.6%, 1.7%
Canada — 1.8%, 1.5%

2005
2018

(%) share of global emissions

Figure 2.1. Greenhouse gas emissions for the world and top ten emitting countries and regions, 2005 and 2018.

Source: Canada, "Canadian Environmental Sustainability Indicators: Global Greenhouse Gas Emissions (2021)," last modified 15 April 2021, https://www.canada.ca/en/environment-climate-change/services/environmental-indicators/global-greenhouse-gas-emissions.html.

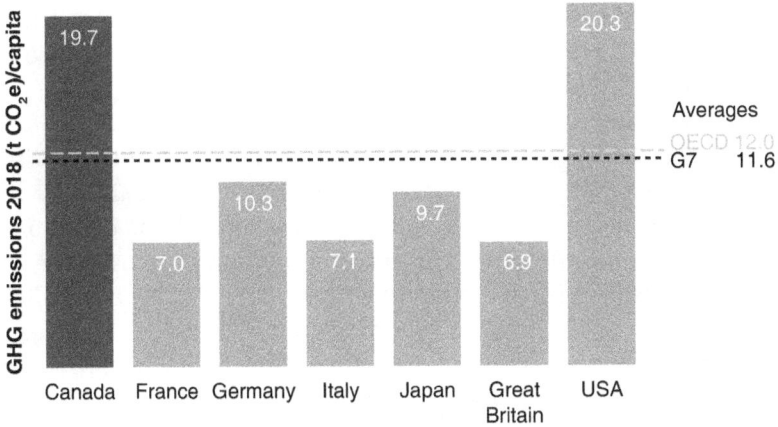

Figure 2.2. Per capita emissions 2018 – G7 countries.
Source: Nichole Dusyk, Isabelle Turcotte, Thomas Gunton, Josha MacNab, Sarah McBain, Noe Penney, Julianne Pickrell-Barr, and Myfannwy Pope, *All Hands on Deck: An Assessment of Provincial, Territorial and Federal Readiness to Deliver a Safe Climate* (Pembina Institute, July 2021), fig. 2a.

Agency estimates that global energy demand fell by almost 4% in the first part of 2020, with coal dropping steeply because of the decline in demand in China.[13] Overall global CO_2 emissions fell by 5.8% in 2020 compared to 2019. However, the IEA forecasted that emissions increases of almost 5% in 2021, driven largely by coal demand, would erase most of that decline.[14]

Canada occupies an interesting position. It is tenth in the world in total emissions, but its emission share is dwarfed by countries like China, the United States, and increasingly India (figure 2.1). However, figure 2.2 shows that in 2018 Canada was second on the list in terms of how much we emit per capita among G7 countries. Moreover, Canada emits well over the OECD average per capita and over twice the GHGs as the average person in G20 countries.[15]

On a global level, climate change poses significant risks to health, food security, water supply, and economic growth.[16] The increased temperature is related to increased extreme weather events and all that goes with them, including risk to humans and destruction of infrastructure.[17] People will be displaced.[18] Species will be lost.[19] People will work less due to high temperatures, leading to large economic losses.[20] More people will be forced into poverty.[21] Women will be particularly harmed.[22]

By one estimate, a 2 degree C increase will lead to a loss of 13% of the world's GDP.[23] A US government assessment noted that "with continued growth in emissions at historic rates, annual losses in some economic sectors are projected to reach hundreds of billions of dollars by the end of the century – more than the current gross domestic product (GDP) of many U.S. states."[24] A UN report cites one estimate that "limiting warming to 1.5 [degrees] C instead of 2 [degrees] C would save 1.5–2.0% of the gross world product (GWP) by mid-century and 3.5% of the GWP by end-of-century."[25] Even with current climate policies, the physical impacts of climate change could reduce global GDP by about 5%.[26]

And for Canada? Canada's National Assessment Process report on the effects of climate change notes that the nation faces "more extreme heat, less extreme cold, longer growing seasons, shorter snow and ice cover seasons, earlier spring peak streamflow, thinning glaciers, thawing permafrost, and rising sea level."[27] Figure 2.3, produced by an expert panel commissioned by the Treasury Board, shows some of the risks.[28] Areas of greatest concern were physical infrastructure, coastal communities, northern communities, human health, ecosystems, and fisheries. Urban areas were particularly at risk, as density increases risk from heatwaves and cities contain a high proportion of vulnerable people.[29] The panel noted that there will be significant but as yet uncertain effects on Indigenous peoples. Similarly, the federal government released a report in 2021 noting that climate change was affecting communities across Canada and reducing benefits from ecosystems such as providing flood control. It forecasted rising economic costs from climate change and increasing costs for adapting to its effects.[30]

We also do not have great estimates of the cost of climate change for the Canadian economy. The Canadian Institute for Climate Choices reported that both the number of and losses from weather-related disasters have risen since the 1970s, with losses "rising from $8.3 million per event in the 1970s to an average of $112 million between 2010 and 2019."[31] They note that many of the coming

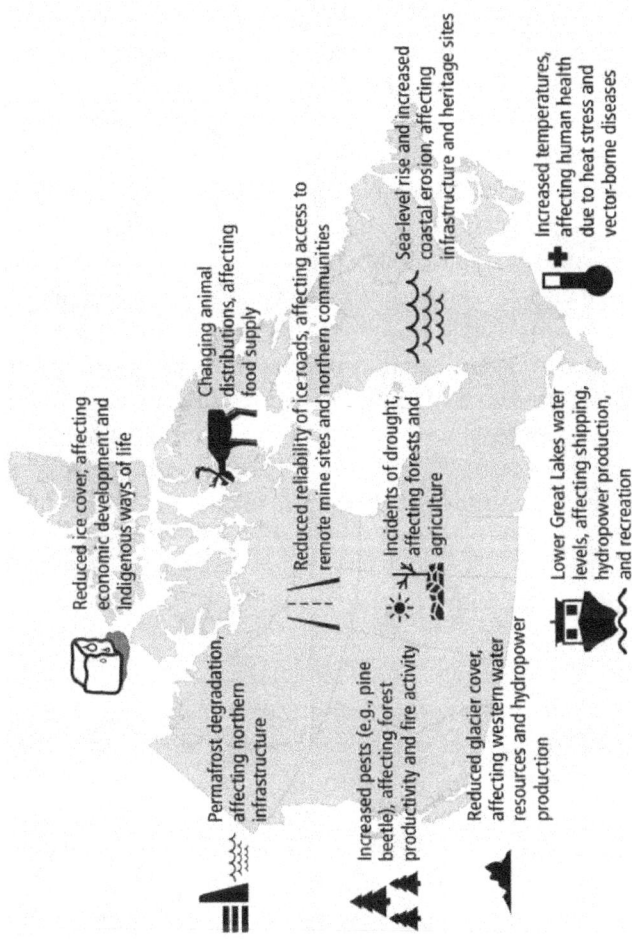

Figure 2.3. Negative impacts of climate change in Canada.

Source: Council of Canadian Academies, *Canada's Top Climate Change Risks* (Ottawa: Council of Canadian Academies, 2019), fig. 1.1.

Adapted with permission from GC (2014)

costs of climate change have not yet been assessed for Canada but that "a changing climate is impairing prosperity and well-being in Canada through economic, social and environmental impacts."[32]

To limit temperature increases to the Paris Agreement target of 2 degrees C, the world needs a 25% decline in emissions from 2010 levels by 2030, and a 45% decrease to reach the stretch goal of 1.5 degrees C.[33] By 2050, globally we need to be "net zero" in emissions on one estimate (that is, zero emissions if you take into account the GHG we put into and take out of the atmosphere), or reduced by 40–70% globally on another.[34]

The Paris Agreement allows countries to set their own emissions reductions targets. We will not meet the 2 degree goal with the Paris Agreement commitments countries have adopted. The UN's *Emissions Gap Report* shows the differences between what countries are committing to do and what they need to do to limit temperature increases. It notes that there is a "vast discrepancy" between the goals and what countries had committed to do in 2020.[35] For just the 2 degree C goal, countries needed to triple their targets for global emissions reductions.[36] According to the United Nations, we need "urgent action at an unprecedented scale"[37] – the next decade is decisive.[38] Some countries including Canada strengthened their targets in 2021, but more is needed.

Climate Change Policy in Canada: A Very Brief Primer

What have we done in Canada? Some, but not enough. Before we look at why this might be and what we should do about it, it is helpful to go through a basic history of climate policy in Canada. Figure 2.4 gives the essence of the story. It shows the targets Canada has set over time for GHG emissions, and the actual and projected emissions. It measures emissions in megatonnes (Mt); a megatonne is one million metric tonnes and is equivalent to the emissions of 250,000 cars in one year.[39]

The two largest sources of GHG emissions in Canada are the oil and gas industry (about 26% of emissions in 2019) and

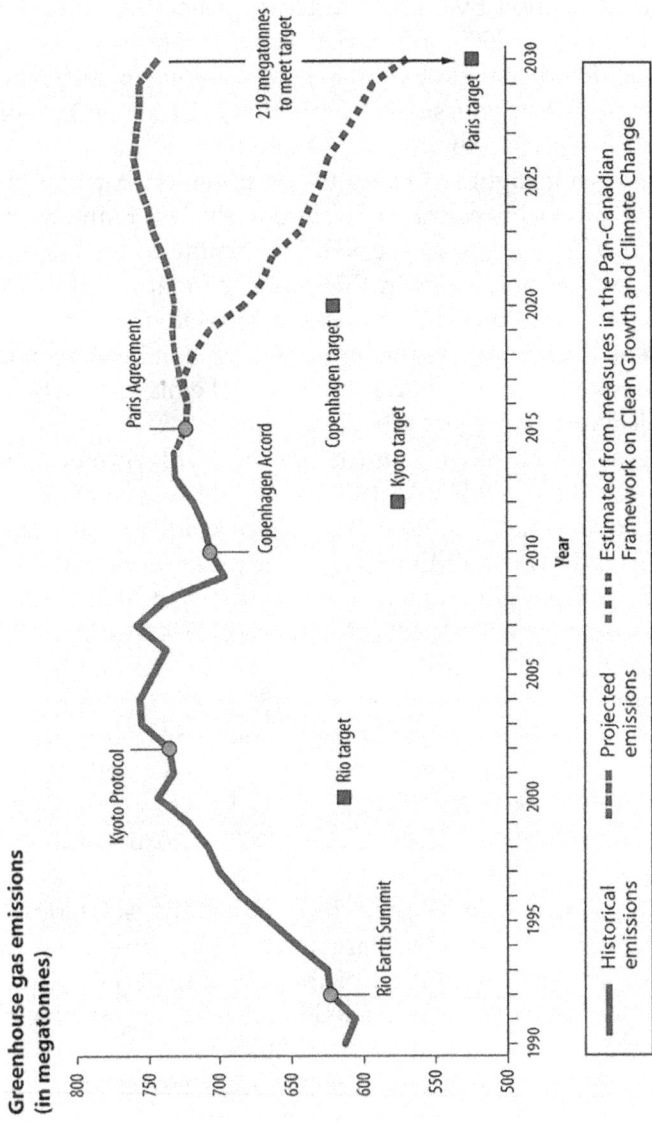

Figure 2.4. Historic GHG targets and emissions in Canada.

Source: Auditor General of Canada, *Reports of the Commissioner of the Environment and Sustainable Development to the Parliament of Canada: Fall 2017: Report 1: Progress on Reducing Greenhouse Gases: Environment and Climate Change Canada: Independent Auditor's Report* (Ottawa: Auditor General of Canada, 2017), Exhibit 1.5.

transportation (about 25.5% of 2019 emissions).[40] Not only are they the largest, but both increased from 2005 to 2019 (oil and gas by 20% and transportation by 16%). Electricity generation was the third largest source in 2005 but, in part due to Ontario's ceasing the use of coal, fell by almost half to 8% of emissions in 2019. The other major sources of emissions are buildings (at about 12% of 2019 emissions), heavy industry (10.5%), and agriculture (10%).

Canada first set a slight reduction target at the Rio Earth Summit in 1992. It joined 154 countries in signing the UN Framework Convention on Climate Change (UNFCCC) aiming to "stabiliz[e] greenhouse gas concentrations in the atmosphere at a level that would prevent dangerous anthropogenic interference with the climate system."[41] Canada's reduction target was not binding and not met. The solid line in figure 2.4 shows actual emissions, which kept rising, leaving the target far behind.

Canada joined with other countries in the Kyoto Protocol in 2005, which set a target of 576 Mt (6% below 1990 levels) by 2012. It also missed that target, and in fact pulled out of the Kyoto Protocol in 2012 under Prime Minister Harper's Conservative government. The federal government next set a target at the 2010 Copenhagen conference of 620 Mt (17% below 2005 levels) by 2020. It abandoned that goal by 2017. The most recent Canadian target was set under the Paris Agreement in 2015. The federal government initially committed to reduce emissions to 523 Mt by 2030 (or 30% below 2005 levels). We were not on track to meet that goal with measures in place in 2019.[42] In 2021, the federal government announced a new target of 40–45% below 2005 by 2030 – or in the neighbourhood of 450 Mt.

Our story, then, is that Canada has continually set targets, missed them, and set new, lower targets generally further out in the future. Why? That's a big part of the rest of the book. However, the pattern has been that at the federal level, the government continually makes plans and, as they say, "plans to make plans" – and took very little actual action. It is not a partisan story: both Liberal and Conservative governments failed to act. The federal government had the National Action Plan on Climate Change (1995), Action Plan 2000, Project Green (2005), Turning the Corner

(2007), and the Climate Change Action Plan (2010), all across different prime ministers and parties.[43] The plans talked of subsidies, information campaigns, renewable energy and, at times, regulatory limits on emissions by different industries. Yet little changed and, in fact, emissions kept rising. Some action was taken, but not enough to make enough of a difference.[44]

In 2016, the Trudeau government announced the Pan-Canadian Framework. All provinces, except Saskatchewan, initially signed onto this policy. Its centrepiece was a required nation-wide carbon price starting at $10 per tonne and increasing by $10 per year up to $50 in 2022. The federal government said that it would allow each province to figure out how best to put in place this carbon price. If a province failed to do so, the federal government would impose it on the province through a "backstop" price. The Pan-Canadian Framework included other actions such as increasing the amount of "clean fuel" used and investing in innovative technology.

Even if the Pan-Canadian Framework worked exactly as the federal government planned, it would not have been enough to meet the Paris target. Environment and Climate Change Canada estimated there would be a gap of about 77 Mt in 2030 – that is, Canada would fall short of the required emissions reductions by a considerable margin.[45] Canada is not alone in this regard – in 2019 only six of the world's major economies were on track to meet their post-2020 targets (China, European Union, India, Mexico, Russia, and Turkey) while seven, including Canada, were not (the others are Australia, Brazil, Japan, Republic of Korea, South Africa, and the United States).[46]

Not everything worked out smoothly under the framework, however. The provinces had previously made some efforts to address climate change, and in fact some took quite impressive action. For example, BC implemented a carbon tax in 2008 that is seen as a leading climate policy internationally. Ontario eliminated its coal-fired electricity plants, making huge strides in addressing both climate change and air pollution more generally. However, in 2018 more than half the provinces did not have GHG targets.[47] Moreover, many provinces had high-level plans but no precise idea of how to implement them.

While the Trudeau government's Pan-Canadian Framework seemed at first to be a Canada-wide agreement on climate action, any semblance of consensus quickly disappeared. When the framework was adopted, it required provinces to have a carbon price in place. Not coincidentally, three of the large provinces already had one (BC, Alberta, and Quebec) and Ontario was in the process of putting one in place through a cap-and-trade program. Saskatchewan was the only province or territory that did not sign on. However, the time since has been stormy. Ontario elected a Conservative government that cancelled the cap-and-trade program. Albertans also elected a Conservative government that cut back some of their climate policies although it retained a carbon price. The federal government introduced its "backstop" carbon price on provinces without one, but Saskatchewan, Ontario, and Alberta launched constitutional challenges to this backstop price that were ultimately unsuccessful before the Supreme Court of Canada.[48]

By 2019 emissions had barely budged, dropping only about 1% from 2005 levels.[49] In December 2020 the minority federal Liberal government announced a new plan, *A Healthy Environment and A Health Economy*. It included a significantly increased carbon price and other measures leaning heavily on spending, which the government estimates will enable Canada to meet its Paris commitments.[50] It was, however, a plan, not legislation, and a plan by a minority government. As we discussed, the federal government subsequently attempted to bake in these reductions by strengthening its Paris commitments to 40–45% below 2005 levels by 2030.

Despite the increased attention to climate change, concerns about Canadian climate policy persist. The Trottier Institute, for example, stated that the federal and provincial plans remained "grossly insufficient to turn the trend in GHG emissions," with emissions declining only to 13% below 2005 levels even with the increased carbon price.[51] The Pembina Institute noted that "across Canada there are glaring gaps in policy infrastructure necessary to achieve climate success."[52] Much remains to be done. The federal government's newest plan, the 2030 Emissions Reduction Plan, adds more spending and promises of regulations in the hope of overcoming some of these concerns.

So far we have looked at the two big levels of government – federal and provincial. In Canada, however, we have other important governments. We will talk about the role of Indigenous governments and Indigenous peoples more generally throughout the book. Indigenous communities are facing the effects of climate change and calling for action. How they are treated within the existing institutions and how we bring about change in a manner that fosters reconciliation will be essential elements of the way forward. What about the local level? Municipalities have implemented many interesting and innovative ideas to reduce emissions and play an important role in such activities as urban planning, transit, and bicycle use. Other non-government actors such as conservation groups and citizens groups have decided to act in the face of inaction by the governments. While these are important actors, in this book I will consider mainly the higher orders of governments, as the scale of action required means these governments are going to have to play a central role.

Of Cows and Climate Change

To understand what has happened and what we need to do, we have to briefly talk about the structure of the problem. Climate change is a "commons problem," a concept explained by ecologist Garrett Hardin in the 1960s.[53] Imagine you live in a town with a public pasture or "commons." You want to get the benefit of grazing your cow on the public commons, but so does everyone else. For every animal you put out to graze, you get the benefit (from its milk or by selling it) but bear only a small part of the cost. What is the cost? It's the impact on the commons of that cow grazing, which is shared with every other user. The problem arises because everyone sees the benefit from putting out another cow but only feels part of the cost – and the result is too many cows. There is over-grazing and the commons becomes less useful or even useless to everyone. Hardin stated, "Ruin is the destination toward which all men rush, each pursuing his own best interest in a society that believes in the freedom of the commons. Freedom in a commons brings ruin to all."[54]

There are commons everywhere – fishing in the ocean, water pollution, a kitchen in a student house. The atmosphere is a commons, as we get the benefit of emitting GHGs into the atmosphere but bear only a small share of the costs, if any at all. The costs of climate change are spread across everyone now living in the world and extend to future generations. Each person acting rationally to further her own interests leads to the problem for everyone. Economists also talk about climate change as a "global public good" or a "collective action problem." The core idea is the same.

The flip side is that we all benefit from a lower risk of climate change if GHG emissions are reduced, whether or not we personally do the reducing.[55] We would prefer to sit back and keep emitting while others lower their emissions. This is called "free-riding": we get to enjoy our cars and our burgers while someone else bears the burden of saving the climate. The difficulty is how to get people to act together and bear costs when you cannot exclude anyone from the benefits of a better climate.

There are different types of solutions to commons problems, which we'll talk more about throughout the book. One is that some authority steps in and tells everyone what to do. The mayor may be able to tell people how many cows they can graze on the town's commons. At the national level, the government may be able to address climate change by setting limits on GHG emissions or taxing or limiting certain activities to ensure we do not overuse the common atmosphere. The difficulty in the case of climate change is that at the international level, there is no overarching authority that can tell all the countries what to do. Hardin spoke of the need for "mutual coercion" – agreeing on limits on the use of the commons. This is, in effect, what international agreements attempt to do, so far unsuccessfully.

Alternatively, we could privatize the commons. The theory goes that if you owned the field that was once the "commons," you would reap the benefits of adding cows but would also bear all the costs of overgrazing. You would not be able to graze as many cows in the future if you ran the field down, so you take care to avoid over-grazing. Whether directly or indirectly, privatization is central to market-based solutions to environmental problems.

There is a third solution to commons problems – people may follow a norm of "appropriate" behaviour, such as the proper number of cows to put on the commons. It is a norm of cooperating. Elinor Ostrom, the first woman to win the Nobel Prize in Economics, found that local commons – such as local fishing areas or pasturelands – did not always end in tragedy. She studied different situations where there should be no cooperation according to economic theory but in fact people were able to act together to ensure a good outcome. Norms were central to these successful outcomes, as they will be to our story of possible ways forward on climate change.

Climate change is so difficult that some call it a "super-wicked problem,"[56] a kind of problem that has four features. First, "time is running out," as the risk of damage increases while we wait to make changes. Second, each one of us is both the cause of and the solution to the problem. We all emit GHGs and we all need to be part of the solution to reduce them. Third, no single government (or other central body) can control all of the choices that lead to the problem. Finally, people tend to care more about short-term losses or gains than the long-term impacts of the problem. Super-wicked problems may have no single answer but require a range of approaches.

There is one last point about the overall structure of the problem that we will come back to when we talk about various policy options. Climate change is, as we saw, a public goods problem. If you reduce your emissions, everyone is better off as you cannot exclude anyone from sharing in the benefits. But economist Scott Barrett points out that there are different types of public goods problems.[57] Climate change is an "aggregate" public goods problem. The solutions to such problems lie in the total amount of action taken globally. It matters less who reduces emissions than that we reduce them overall by some amount. This is different from other public goods problems like the COVID-19 pandemic. The pandemic was more of a "weakest link" public goods problem – the effectiveness of any solution depends on what the one who does the least is doing. We could potentially self-isolate and slow the transmission, but if others are not doing so it may not matter that we have stayed inside our house for a month. Climate change, then, has a slightly different structure than the pandemic, which will be important to our discussion about the efficiency and fairness of different solutions.

That, then, is the overall problem. The science is telling us that we need to reduce emissions significantly in the short term in order to stave off larger, long-term impacts. These impacts will depend on the total amount of emissions globally. However, we face a commons problem. Everyone would prefer that someone else does all the work while they sit back and reap the benefits of reduced risk of climate change. The structure of the problem is simple but hard to crack. And Canada has its own story that adds to the difficulty.

Our Place in the World

Before we dive into the Canadian experience, we need to address a key question – do Canadian institutions even matter? Canada is the tenth-largest emitter but our emissions are dwarfed by those of the United States, China, or India. We care about the *aggregate* global amount of emissions, so if Canada reduces its emissions but those reductions are offset by an increase in another country, we have stood still. Further, Canada is no different than any other country – it is rational in a narrow sense to prefer that other countries reduce their emissions while Canadians continue to benefit from the production and use of fossil fuels for a while longer. We have heard as much recently from prime ministers, both Conservative and Liberal.[58]

It looks like emissions reductions in Canada will be pointless within the bigger picture and may only hurt the Canadian economy. But is that true? There are at least three counterarguments, one moral and two pragmatic. The moral reason for reducing emissions comes in different forms.[59] Some rights-based views start with the idea that everyone is equal and on that basis would allocate emissions on a per capita basis.[60] On this view, Canada would get only a small proportion of future emissions, much smaller than its current per capita emissions. A second rights-based view focuses on the idea that those who have resources have the right to use them or, at very least, a country's share of emissions should depend on what their emissions levels have been in the past. If we need a 30% reduction in global emissions, everyone cuts their emissions by 30%. This view is closest to where Canada stands now. Third, a corrective justice approach would lay responsibility on whoever caused the problem. What matters for climate change is not merely the emissions in each year but the amount, or stock, of GHGs that is already in the atmosphere. Importantly, it stays there for up to a thousand years.[61] The corrective justice view would ask countries which that contributed the most to the stock of GHGs to bear more of the costs of reductions.[62] Canada, on this view, should bear costs of reducing emissions because it has historically contributed significantly to the stock of GHGs – about 2% of the cumulative GHG in the atmosphere.[63] Finally, there are other moral arguments, such as those based on fairness.[64]

We could debate which of these views holds the most power, but on any of them Canada would emit less than it does now. Even the proportionate share notion leads to Canada taking on some reductions. There is, of course, a difference in the degree to which it would act, but none of these approaches counsels disavowing any responsibility. As far back as 1990 the federal Green Plan recognized this, acknowledging Canada was a small total emitter in global terms, but stating that "we should not allow our relatively small emissions to be an excuse for not taking action at home." [65] And Canada is not that small – it is a top-ten emitter globally.

Second, even if you don't buy into the moral argument for Canada to act, climate change could be seen as an "opportunity" for Canada to grow economically. In fact, this framing is key for the federal and many provincial climate plans. A recent federal government plan was entitled *A Healthy Environment and a Healthy Economy: Canada's Strengthened Climate Plan to Create Jobs and Support People, Communities and the Planet*.[66] Former minister of the environment and climate change Catherine McKenna stated that "the clean-growth economy is where the world is going and

Canada is going to be part of it … Canada wants to create good jobs, and wants to grow our economy and create opportunities for business."[67] Many of the provinces also seek this magic combination. The federal government anticipates massive gains, stating that "bold action on the climate could yield direct economic gains of US$26 trillion, compared to business as usual" and create as many as twenty-four million jobs globally by 2030.[68] The International Renewable Energy Agency estimates that investments in renewable energy, and improved energy efficiency, could exceed $95 trillion between 2016 and 2050.[69]

Business leaders, environmental groups, and governments all see opportunity in this huge market. It would enable Canada to lean less heavily on fossil fuels. The worry is that Canada could be left behind as other countries also try to tap into this market, particularly the Biden administration in the United States. Luckily, Canada seems to be starting from a strong base of expertise in energy. On one ranking, Canada was fourth in the world in green innovation in 2017.[70] Moreover, the Canadian Institute for Climate Choices notes that Canada is "well positioned to capture enormous market opportunities from the global transition," in part because it is in the top ten countries for producing or holding reserves for minerals necessary for the transition.[71]

One important reason for climate action, then, comes from the economic possibilities. The federal government's climate plans pick up these themes, proposing money to build Canada's "Clean Industrial Advantage" to spur new technologies to decarbonize existing industry.[72] The question on this view is not whether a transition in the Canadian economy will occur, but how fast it will occur and how it can be used to boost growth. "The green wave is coming," according to the Canadian Institute for Climate Choices, but Canada is not sufficiently prepared.[73]

For those who do not buy the moral argument or are worried about Canada's ability to become a leader in the green economy, there is a third argument that it is in Canada's best interests to act now: Canada is necessary, though not sufficient, to the fight against climate change. If Canada fails to meet its targets, its emissions on their own will not take the world over the edge. But again, climate change is a commons problem, and one solution to such problems is cooperation based on trust. Ostrom noted that to increase collective action, you must "enhance the level of trust by participants that others are complying with the policy or else many will seek ways of avoiding compliance."[74]

Trust is tricky – hard to build, easy to lose. You have to build trust over time and it requires action – once you show that you are trustworthy by complying with a norm, others will be more willing to follow through as well.[75] It can, of course, go the other way – showing that you will not act makes it less likely others will act.[76] We will look more into the elements of building trust later, but they can be difficult to meet – there must be some agreed-upon norm, information about what each party is doing, some means of sanctioning those not acting as they should.

Ostrom's approach gives us insight into international action.[77] While we will discuss in chapter 4 how some feel greater incentives for international action may be needed,[78] action by Canada may be necessary for cooperation in the international regime. By not acting, Canada gives others license to do the same.[79] It gives cover to both developed and developing nations to free-ride on the hoped-for reductions by others. Acting, on the other hand, supports action by bigger players, fostering the sense that everyone will play a role. Canada can help bring about a more positive, fairer outcome. We should not get carried away with how important Canada's role is, but neither should we neglect the effect it can have on the outcome.

Fortunately, you do not need to believe all three of these arguments to make the case for Canada to act. Any one will do to ensure Canada plays a role. If Canada does nothing, it increases the likelihood that international cooperation will fail, it will miss the opportunity that climate change provides to take advantage of the coming changes, and it will have failed in a moral obligation.

"There Was a Time in This Fair Land When the Railroad Did Not Run ..."

So begins Gordon Lightfoot's "Canadian Railroad Trilogy" about the growth of the railroad in Canada and how people came from other countries and "built the mines, the mills and the factories for the good of us all." The way Canada grew after Europeans came is important for both the nature of the problem we face in addressing climate change and the potential solutions for Canada. It frames the government's relationship with Indigenous peoples and how Canadian political and economic structures developed. Everything we saw about the commons problem applies to every country. We know the basic problem and the general solutions. The challenge, and the only hope, is to think about how these translate into the Canadian context.

The key to understanding this context goes back to an important Canadian contribution to the theory of how countries grow economically. It is called the staples thesis, associated with Canadian economist Harold Innis, who in the 1930s studied how natural resources like fur and fish played a role in Canada's economic development.[80] The idea is that countries like Canada use their natural resources (staples) to kick off growth. They slake the demand from other countries for these staples – the countries in the "centre" extract what they need from the hinterlands or margins such as Canada. In theory, these staples economies can use the wealth from these natural resources to transform the economy in other directions.

At the same time, being dependent on a staple has downsides. A country can deplete its staple without replacing it with a longer-term basis for the economy. Staples are also subject to booms and busts, which can potentially lead to instability and inequality. Further, being so focused on one sector can hinder other potential exports. Countries can fall into a "staples trap": a poorly diversified economy with other sectors not developed enough to be relied on if things go wrong with the natural resource sector.[81] While Innis saw the risks, others viewed natural resources as fostering export-led growth, leading to development in other sectors.[82]

Canada started as a nation of "hewers of wood and drawers of water" – relying on fur, wood, and then wheat for economic growth. We have moved from staple to staple, often depleting as we go. The decimation of the beaver and the bison in the early years of colonialization came from competition and lack of control over their harvesting.[83] Canada has also done poorly in protecting its fisheries. The collapse of the cod industry despite government regulation is the most spectacular example, but by no means the only one. Canada subsequently also developed its non-renewable resources such as minerals. While the notion of depletion is different (minerals, for example, by their nature are not renewable in any realistic time frame), the underlying structure of the issue is the same – Canada became dependent on exploiting these resources to further growth.

Canada's natural resources tend to be on public lands. Much of its forestry and mining, for example, occurs on publicly owned lands – mostly provincial. Governments use various forms of leases to grant access and use to private companies. These industries bring jobs and economic growth as well as government revenue, both directly in the form of payments such as royalties or stumpage fees and indirectly in the form of corporate or personal income tax.

Some argue Canada is similarly now relying on a new "staple" – oil and gas (and to a lesser extent coal).[84] In 2018 "Canadian exports of petroleum products topped $125 billion; that is 65% higher than auto exports, and nearly triple that of base metals (the closest resource-based comparable)."[85] Oil and gas contributes over 7% of Canadian GDP and accounts for 20% of its trade.[86] Canada is a major exporter of fossil fuels – fourth in the world in exports of oil.[87] And it has plenty of oil left – Canada has the third largest proven reserves of oil, 96% of which are located in the oil sands.[88] Coal is BC's top export, natural gas its fourth largest.

This raises a complex question for Canada: "is natural capital wealth a 'curse' that constrains long-term growth or a springboard to a resilient economy that can sustain growth?"[89] Natural resources have on balance had positive economic effects for Canada, though the impacts seem modest compared to other factors such as infrastructure and machinery or our people and institutions.[90] Economic growth in the 1990s and 2000s seems to have depended on Canada's substituting natural resources for these other sources of growth.[91]

This history of reliance on resources has two big implications for our discussion of how Canada has addressed climate change and how we should think about reforms. First, and most obviously, the fossil fuel industry (and use of fossil fuels) is very important to the Canadian economy, which elevates the costs of climate action. We need to consider the loss of economic growth, the costs in terms of lost jobs, the investments that will become less productive or obsolete, the need to change infrastructure. These costs make what is a difficult transition for all countries particularly challenging in Canada.

There is, though, a second concern. For political scientist Brendan Haley, the difficulty is that the dependence on oil and gas leads to institutions and interests that further our dependence on the oil and gas industry.[92] Haley sees the reliance on export-led growth as baked into Canada's institutional DNA – Canada is predisposed to making things work for an economy based on oil exports because of its history with fur, wood, and wheat.

Canadian institutions shape Canada's use of its natural resources – through rules about harvesting rates or forms of dealing with pollution from mines, through rights of ownership over the resources, through the fees Canadian governments charge for harvesting or mining. At the same time, Canadian natural resources have shaped its laws and institutions. Harold Demsetz, an economist, argued that when the underlying costs and benefits of exploiting a resource change, the property rights change. He suggested that the demand for fur changed how Indigenous groups set up their hunting territories. However, the effect is not straightforward. Politics also shapes how institutions develop – whether towards or away from efficiency or fairness.[93] Canada's dependence on natural resources has shaped its institutions, including the form and substance of environmental laws, the relationship of the government to Indigenous peoples, and the Constitution itself.[94] It even, according to legal scholar Jason McLean, limits our imagination concerning what are possible solutions.[95] It hampers the ability to pursue technological or economic innovation that would allow Canada to transition away from this dependence. In short, it pushes Canada towards a "carbon trap."

We will talk more about how reliance on oil and gas may influence Canadian institutions and political choices. For now what is important is that Canada has long been an economy heavily, though not solely, reliant on trade in its natural resources. Currently oil and gas play a large role in its economy. Yet it looks like the global economy is moving away from fossil fuels; by some estimates the world will hit peak oil demand by the 2030s, possibly sooner, though economist Mark Jaccard doubts that "peak oil" will be important given the amount of remaining supplies.[96] Renewables (including nuclear, biomass, and hydro) could serve 25% of the global energy demand by 2035, rising to 34% by 2050.[97]

There is no necessary connection between a resource-based economy and unsustainable exploitation, but strong institutions are central to a more positive story. To understand how Canada as a country can address climate change, we need to keep in mind this complex interplay between our natural resources, economic growth, and institutions. Canadian institutions structure how natural resources are used through how they distribute power – to the federal government versus the provincial governments, for instance, or to markets versus the government more broadly. However, natural resources have at the same time structured Canadian institutions – its Constitution, laws, redistribution policies, relationships with Indigenous peoples. We will not achieve a sustainable Canadian climate policy without recognizing this connection to our historical and natural context.

chapter three

Discretion I: Picking the Wrong Tools

Carbon tax. Carbon price. Carbon levy. Whatever you call it, it is all over the news these days. The federal government is touting its carbon price as essential to the fight against climate change. The Supreme Court of Canada has held that in order to ensure Canada addresses climate change as a nation, the federal government has the ability to impose a minimum carbon price on reluctant provinces.[1] Economists proclaim a carbon price to be the best weapon for battling excessive GHG emissions. The federal Conservative party has recently even proposed its own carbon pricing plan.

At the same time, some politicians are claiming that the federal carbon tax is not doing anything to fight climate change but instead is destroying the economy and harming working families; the cost is too high given the risks. In its place they argue that we should give money to families to change their ways. They point out that it is not that they do not believe climate change is happening; it is just a disagreement over the best way to confront it. Move too quickly, they argue, and we are imposing unnecessary costs on ourselves.

One good thing about this debate has been that it represents widespread acknowledgment that something needs to be done. The downside is that the focus on the price has both distorted the role that price is playing and deflected the debate from what we know about the different ways of reducing the risk. Market-based policies such as a carbon tax or a cap-and-trade system are relative newcomers to environmental law generally, let alone climate change law. The more traditional approach has been to use various forms of regulation such as bans or limits on various types of emissions or, alternatively, subsidies or voluntary programs to help individuals or companies take more environmentally friendly action.

While we have some experience generally with how to reduce environmental risks, climate change is different. The changes needed are so large that it is hard to estimate the costs given that we have not undertaken anything on this scale before. Moreover, we do not have experience with catastrophe resulting from a rise in temperatures that would allow us to estimate the risks. We are, in economists' terms, in the area of uncertainty as opposed to risk. Something is a "risk" if we can with some accuracy predict the probability of a resulting event occurring, an "uncertainty" if we cannot. While science has given clear probabilities of some of the harms from climate change, we do not know enough to accurately assign a probability of complete catastrophe from climate change.

Given this uncertainty, how can we decide what steps to take? It turns out we know a lot about what tools we can use to reduce emissions; it is just we do not use them effectively.

What Do We Want?

We saw in the previous chapter that we know the basic ways to address a commons problem. So why is there so much disagreement over which tools to use to fight climate change? Part of the reason is that some do not want to engage in the fight. They say anything we do is too expensive and is pointless. However, there is more to the issue than that. Canadians generally agree that climate change is an issue we need to do something about. We just do not seem to be able to agree on what we want to achieve and whether a particular policy can actually achieve it.

There are three broad ways to assess different policy options: efficiency, fairness, and feasibility.[2]

Efficiency. If you asked an economist what steps to take to reduce the risk of a fire, she might say, "Tell me what level of risk you are willing to bear and I'll help you find how to reach it at the least cost." She would be using one version of "efficiency" to decide what to do – reach whatever target you want in the most cost-effective manner possible. In the climate context, we would set the level of emissions reductions we want and then try to find the least costly means of reaching it. You can imagine any number of ways to reduce emissions to a particular level: We could immediately prohibit all GHG emissions, but the cost to the economy would be extraordinarily high. We could phase in reductions, but we then continue to build up the stock of GHGs in the atmosphere, possibly making it harder to reach our goals. We will need to think about the costs of complying with the rules in the short term, but also about whether they provide incentives for individuals to change behaviour in the long run.

"Efficiency," though, is often also used in a broader fashion – of not just hitting a level of risk at the least cost but also of choosing that level of risk. "Efficiency" in this sense is about trying to meet all our societal goals, such as dealing with inequality, improving health, and providing education as well as reducing the risks from climate change, with the resources at our disposal.[3] How do we choose what risk of catastrophe we are willing to live with? This raises the issue of "opportunity cost" – thinking not only about the cost of reducing a risk, but also the value of that action relative to other things you could do. In the climate context, we would need to think not just about how much it costs to use a tax to reduce emissions, but also how much we value the reduction in emissions relative to other things like reducing poverty. When we talk about the efficiency of climate policy, we need to be careful about whether we mean this narrowly in terms of hitting a target at least cost or more broadly as trying to set our societal goals.

Fairness. Policies can also be evaluated based on their effects on the lives of individuals – who is benefiting or being harmed? The fairness of these effects will obviously be contentious. We will discuss the issue of fairness more in chapters 7 and 9; for now we can say that all tools will impose a cost, but different tools may imply different individuals face higher or lower costs from climate action.

Feasibility. Finally, we may be able to design clever ways to reduce emissions, but can we actually make them happen? Because different policies impose costs and benefits on different groups, the likelihood of these policies being implemented will vary. We need to take into account the feasibility of getting policy adopted, not just whether we can design the perfect plan.

Comparing Tools

There are four broad types of tools we can use to reduce GHG emissions: carbon pricing, regulations, subsidies, and other measures. How do they stack up? We will look at efficiency and feasibility and leave fairness for later.

Carbon Price (Market Mechanisms). Economists love carbon prices. The idea is intuitive. Climate change is a commons problem – we get the benefit of emitting GHG but do not pay the costs and so emit too much. Make something more expensive and people will try to find ways to avoid paying – either by doing it less or by changing how they do it.[4] Increase the cost of gas and people will either drive less or buy cars that use less fuel. Increase the cost of electricity and people will turn off lights or start using more efficient refrigerators.

You can price GHG emissions in two main ways – a "tax" or a cap-and-trade mechanism. A tax is just a direct price on emissions. Under the current federal system, it costs $40 to emit a tonne of CO_2. It is of course not that simple, but that is the core idea. A cap-and-trade system is a bit more complicated. The government decides how many tonnes of GHG it will allow to be emitted in a given year. It then divides this total amount into smaller units – say, one tonne of CO_2. The government can either give these one-tonne "allowances" away to companies or sell them; either way is efficient. The company then decides how much to emit based on these allowances. For example, a company that receives ten allowances can emit ten tonnes of CO_2 without penalty. If it emits more than ten tonnes, it has to buy extra allowances from some other company that is willing to sell to them. If it emits less, it can sell its excess allowances to someone who needs them. Either way the company faces a price – either of buying allowances or of not selling allowances it could sell if it reduced its emissions.

From an efficiency point of view, both means of pricing are essentially identical. The beauty of using a price to tackle climate change is that the government does not have to decide the best way to reduce emissions; the price for each tonne of GHG emitted pushes people and businesses to think about ways to reduce emissions. Is it cheaper to buy a smaller car or drive less? Can we make cars that emit less? The price makes people think about these questions and unlocks their creativity. It provides incentives for innovation. And it can give the government money to do good things with – reduce other taxes, fund other emissions reductions or, as we will discuss, alleviate the negative effects on individuals or businesses.

But there are differences. With a carbon tax you know the price of a tonne of emissions but not what amount everyone will ultimately emit. A cap-and-trade system is the opposite. It gives some certainty about how much will be emitted but the price is uncertain.[5] As the price will vary depending on the number of tonnes the government allows to be emitted and companies' demand for those allowances, it can fluctuate wildly.[6]

Further, a tax seems easier to implement. The government could set a price and administer it through the general tax system. A cap-and-trade system requires much more infrastructure. The government has to create the allowances, provide for market rules, figure out a way to distribute the allowances to all those who may need them, monitor the emissions, and enforce the rules. For taxes, though, things can get complicated quickly. It is hard to know the right tax rate. Set it too low and you get too high a level of GHG emissions; set it too high and you risk discouraging people from doing things that we otherwise want them to do. For economists the goal is to set the price at the level of harm caused by the emissions so that the emitter faces the full cost of its actions. If it costs $100 to produce a tonne of steel but the GHG emissions create $25 of damage, you want the steel producer to take into account not only the $100 but the $25 in order for it to make the efficient choice about how much steel to produce. We will talk about the difficulties of estimating this damage (called the social cost of carbon) later in the chapter.

The biggest downside of pricing relates to feasibility. As economist Mark Jaccard points out, many economists fail to take into account the level of pricing that is politically possible.[7] Pricing imposes a visible cost on people and companies, and so it is hard to put in place and even harder to make the price high enough to make a difference – the level of the carbon tax may need to be very high to significantly reduce emissions. The Ecofiscal Commission estimates that Canada needs to impose a carbon price of $210 per tonne by 2030 just to meet the initial targets it set for itself under the Paris Agreement.[8] Carbon prices, though, have gained some support from business leaders,[9] in part because pricing brings some certainty for businesses.

Carbon taxes are an important part of the answer but will not be enough. Nobel-prize- winning economist William Nordhaus argues that to effectively fight climate change people need to understand the seriousness of climate change and countries need to increase the carbon price.[10] These conditions are obviously connected. You won't get the second without the first – and Canada does not yet fully have the first.

Regulations. Regulations can take many different forms. In the extreme, they can prohibit something, such as banning the sale of all gasoline-powered cars. They often, however, take a less extreme form, such as requiring all car manufacturers to install a particular device that reduces emissions. Still more flexibly, they can mandate some level of performance, such as requiring that car manufacturers ensure all cars meet a particular standard of kilometres per litre of gas.

Regulations generally cost more for a given level of emissions reduction than does imposing a carbon price.[11] Regulations that specify a particular measure to be taken are the most expensive.[12] They are costly to design, as the government has to decide which measures are possible and most effective. They should also adjust as technology changes, and so the government has to monitor each area it is trying to regulate. Not only is it expensive to do all this, it poses a large risk of the government getting it wrong. It can also blunt incentives to improve – once a company puts in place the required emissions-reduction device on your car, it is done. It does not need to think of better ways to reduce emissions.

However, regulations need not be so blunt or so expensive. Jaccard argues that you can design "flexible" regulations that may not be as efficient as pricing but are close.[13] Such regulations, for example, set a standard that has to be hit – a level of emissions, a percentage of cars that have low or zero emissions, etc. There is flexibility about how to meet that standard – what emissions reduction device or production process to use or what types of cars to produce – that brings with it the ability to keep costs low as well as to innovate in meeting the standard.

The value of regulations over carbon pricing is that the full cost that is being imposed is not transparent – having "implicit" costs may increase the likelihood these instruments will be adopted.[14] However, carbon pricing may have a couple of advantages over even flexible regulations, the first of which is to encourage creative thinking. Flexible regulations that set targets (as opposed to requiring specific technology) give incentives to companies to innovate to meet that target. If the government sets targets for the average fuel efficiency of the fleet of cars sold by each car manufacturer, for example, the car manufacturers would innovate on meeting those targets within the fleet. However, it would not give them an incentive to get out of car manufacturing and into electric bikes. A price, by contrast, may force companies to think of more radical moves. Second, carbon pricing can raise money that may be used to make climate policy fairer by offsetting the negative effects on less-well-off people or struggling businesses (although regulations can also be made more fair in a number of ways, such as through exemptions for certain individuals or businesses).[15]

Subsidies. Subsidies are a less efficient or effective climate tool than pricing or regulations. In theory, governments' subsidizing something should essentially be the same as putting a price on doing the opposite. You can charge people a price for putting gas in their car or you can pay them to drive less. Politicians love subsidies because they get to give money to people as opposed to charge them for something they used to do for free or direct them through regulations to take some costly action.

However, economists argue that governments have difficulty determining what they should pay people to do and how much to pay. The result is often too much money going to ineffective measures. Economists contend governments are no good at "picking winners" – the best type of zero-emission car or renewable energy. They do not have good information about which technology will be the best. And companies spend a lot of money just on trying to convince governments to fund their technology. Resources are used not to help the environment but to win the fight to get the government cheque.

Furthermore, even if we are subsidizing the right things, economists worry that much of the money is wasted. For example, if we subsidize the purchase of electric cars, we may be paying some people to buy electric cars that they would have bought even without the subsidy. The return from the subsidy is often much worse than is claimed. And we have to raise taxes to pay for them, which has their own negative effects on the economy.

Subsidies, then, may have a role in climate policy, but we should not expect too much from them. A closely related tool is for governments to invest in emissions reduction projects. They could, for example, fund more public transit or a network of charging stations for electric vehicles, which can reduce emissions in the transportation sector. These types of expenditures can make things happen that we need but that would not exist, or at least there would not be enough of, if left to the market. Some such large public expenditures will be necessary to foster the transition by reducing the costs of lower carbon options.[16]

Information/Voluntary Measures. The last set of tools takes a step back even from paying people to act. We may feel people do not know enough about the harms from climate change or how to change their behaviour, but if they did they would take more climate-friendly action such as driving less or buying "greener" products. We will have more to say about these types of problems in chapter 11.

Giving people information can be a fairly cheap form of intervention. The difficulty is in deciding what to let people know and how to make them actually care about the information. It does no good to let people know generally about the risks from climate change if they either do not see the risks to themselves or do not care enough about the risks to others. It can be difficult to get large-scale, rapid change through providing information.

Some types of information, however, may make a big difference. Forcing companies to disclose what their emissions levels are or to estimate their vulnerability to the effects of climate change or the costs of climate policy can have a significant effect. Economist Andrew Leach argues that a big source of change for oil sands companies beyond the price of oil is pressure from investors.[17] Some countries are moving to force companies to tell investors about how they are impacted by climate change. Getting information out more widely encourages companies to invest in emissions reductions or new forms of energy. Market forces are an increasingly important lever in reducing emissions that can be enhanced with disclosure of climate-related information.

Finally, another low-cost/low-effectiveness tool is the use of voluntary measures. Governments have in the past set up voluntary challenges under which companies sign up to reduce their emissions by some amount and get recognition for meeting their goals. These challenges are not backed by any formal sanction, but depend on companies caring about their reputation if they do not follow through. They often do not give much of an incentive to reduce emissions as consumers may not punish companies for not meeting their commitments.[18]

A Weak Mix

Each of these four types of tools can make a difference on climate change, though each has weaknesses. Pricing and regulation are the most efficient, though they differ in how politically feasible they are. Subsidies and information tend to be less efficient and effective but more politically feasible. Luckily, the choice does not have to be all or nothing. While people may argue about relative weights for different instruments, all seem to agree that we need a mix.[19] The key to progress is relying mostly on fairly stringent pricing and regulations with subsidies and information playing a supporting role.

Does the choice of tools in Canada look anything like this? The short answer is mostly no, though the mix has significantly improved in recent years. Right now Canadian governments do use a little bit of everything – some subsidies and a little bit of pricing, backed by some (sometimes flexible) regulations – but there have been a few big concerns:

Getting the Right Mix. Much of Canadian climate policy in the past has been the opposite of what the experts would advise.[20] Instead of relying on taxing and regulating, climate policy was built around the less effective and more expensive options of subsidies and information. Governments chose the politically easier task of paying people to reduce emissions. Only recently have they turned to actually imposing a cost – either directly through carbon pricing or more indirectly through regulations.

At the federal level, the early Liberal government plans of prime ministers Jean Chrétien and Paul Martin relied heavily on spending and voluntary action.[21] Conservative prime minister Stephen Harper's government abandoned many of these plans. It embarked on a "sector-by-sector" approach aimed especially at the transportation and electricity sectors, though it failed to have an implementation plan or apparently even rely on economic analysis.[22] As it relied mostly on a few regulations to reduce specific emissions, the measures unsurprisingly did not significantly reduce overall emissions.[23]

The Trudeau government has improved the mix. Its Pan-Canadian Framework included support for green innovation and subsidies for reducing emissions from the oil and gas sector.[24] It introduced a minimum carbon price across the country and proposed more "flexible" regulations, such as a clean fuel standard designed to give scope for meeting an emissions target in a variety of ways. More recently its *Healthy Environment, Healthy Economy* plan built on this use of multiple tools, leaning much more heavily on the carbon price, which they propose increasing to $170 per tonne by 2030.[25] They also propose continuing with a clean fuel standard (though a narrower version) as well as providing subsidies for a wide range of initiatives, including home retrofits, building upgrades, decarbonization of large emitters, clean energy, and zero-emission vehicles. In their *2030 Emissions Reduction Plan*, they further promised to increase the requirement on sales of zero-emission vehicles as well as to place a cap on oil and gas emissions.[26]

Some provinces began pricing earlier. Quebec and Alberta set small carbon taxes starting in 2007, and BC adopted its more famous broad-based carbon tax in 2008.[27] Quebec and Ontario introduced cap-and-trade programs in 2013 and 2017 respectively. The provinces have also used regulations; notably, Ontario took a very large step by shutting down coal-fired electricity generating plants in 2014. Some provinces have used flexible regulations, including BC, which requires a certain percentage of renewable fuels in gasoline; Alberta, which sets emission intensity reduction targets; and Quebec, which specifies the number of zero-emission vehicles that must be sold.[28] The provinces also use subsidies for climate initiatives as part of their carbon pricing plans. The result is that there is now carbon pricing across the country, some of it imposed by the federal government and others adopted voluntarily by the provinces, and most provinces also use some mix of regulations, subsidies, and information.

One final point: much of what we are talking about seems like ex poste action – something is already being done (bitumen being taken from the oil sands or LNG facilities being built) that may harm the environment and we need to take steps to contain it. Obviously, the climate policies we have been discussing may influence whether these things are going to be done. A carbon price or regulations might make a project (a mine or oil project, for example) uneconomic. In a perfect world, the price and/or regulations would be set in a way that would drive individuals and industry to fully consider all the costs of their actions. However, we are a long way from such a world. We will talk in chapter 10 about our processes for thinking about the environmental effects of a project before it is built; it is called environmental assessment (at the federal level it is now called impact assessment).

Getting the Level Right. While Canada has moved towards a better mix of tools, the levels are too low. Clearly the federal carbon price has been too low to meet the Paris target on its own. As we saw, by some estimates it would need to be ten times higher than its current level.[29] A low carbon price can provide some incentive as well as a jumping-off point with the aim of increasing the price over time. Companies and everyone else then will have time to adjust, reducing the costs of acting as well as possibly the opposition to the price. Moreover, the carbon price was never intended to be the sole tool; it was meant to work in conjunction with other policies. Yet even the significant increase in the carbon price promised by the federal government will decrease emissions but not allow Canada to meet its targets.[30]

Still, even with the federal government's proposed increased carbon price, the combined effect of carbon pricing and regulations shows an implicit carbon price that is very low. Moreover, individual policies seem weaker than they should be. Economists would argue we should take measures for which the benefits are at least equal to, if not greater than, the costs. So what is the benefit? For economists, it is the social cost of carbon – the cost in dollar terms from emitting an extra tonne of GHG or, alternatively, the benefit from avoiding the cost of a tonne of emissions. The social cost of carbon is estimated by combining models of the environment with models of the economy (called integrated assessment models). These models are tremendously complicated, relying on multiple equations and many often-controversial assumptions.

Economist William Nordhaus claims these models "play a key role" in designing policy, while Sir Nicholas Stern, another famous climate economist, argues that the models and the cost-benefit analysis tied to the social cost of carbon "constitute a possibly useful but minor part of the relevant evidence" and "are likely to be strongly biased in the direction of weak action."[31] Some governments, including the federal government, undertake a cost-benefit analysis of any new climate policy and assess the benefits of the policy changes using an estimate of the social cost of carbon.[32]

However, governments may be using too low a social cost of carbon, leading to weak policies. In fact, the federal government notes that the assigned social cost of carbon of $50/tCO$_2$ it used in its recent planning "underestimates the damages of climate change to society and the social benefits of reducing carbon pollution."[33] One concern is that the estimates do not fairly take into account the lives of future generations.[34] How much should we be willing to invest today to save lives in the future? The idea is that taking an action (say, to buy electric buses) costs money now but the benefits are in the future – fewer harms to the environment, fewer lives lost or crops damaged. We need a way to bring all those future benefits into today's dollars if we are to directly compare them to today's costs. This is a standard economic question about the appropriate "discount rate," and it can make a huge difference to what we think are the benefits of climate action.[35]

One way to determine the right discount rate may be to look at choices people actually make – on this view, people's investment choices give a measure of how much they are willing to give up today to get more in the future. If we look at people's investment choices in the market, we get a fairly high discount rate – which implies a relatively low benefit from climate action. Stern, on the other hand, maintains we should not use the market rate but should instead think about it ethically, arguing we should not consider future lives less valuable than our own. The result from this "ethical" approach is a much lower discount rate – which implies a greater benefit from action. For 2020, for example, the social cost of carbon jumps from $27 per tonne at a 5% discount rate (approximately the market rate) to $515 at a 1% rate (Stern's ethical rate).[36] The federal government is either explicitly or implicitly using a discount rate that is closer to the higher (market) estimate of the discount rate (and so a lower estimate of the benefits of action).[37]

A second concern is that the calculations of the social cost of carbon fail to adequately account for the uncertainty about the possibility of catastrophe. Economist Martin Weitzman argues that this uncertainty should lead to a higher social cost of carbon, and if the risk of catastrophe is high enough we should be willing to pay an infinite amount to reduce it.[38] While Weitzman does not believe that the risk is likely that high, he argues that it should "serve as a warning flag that a credible economic analysis of climate change should seriously consider" extremely bad outcomes and the probability of their occurring.[39] For Weitzman, this risk is an empirical question that we do not have a handle on, particularly since "climate change is a unique one-off event."[40]

Nordhaus, on the other hand, argues that we can still work within the regular economic models by including the low probability of large harm, as he sees no evidence of an extra risk of catastrophe right now.[41] He refers to climate change as a "vast casino" where we are "rolling the climatic dice" with our future, but he says "there is time to turn around and walk back out of the casino."[42] However, if Weitzman is right, we may not be able to plan our exit by relying solely on regular "marginal" economics, assuming we can safely use the price system to tell us the right amount to invest. We may need to take extra action just in case, as insurance. Or we may need to abandon the attempt to decide the right level of reductions solely through economics and make the big "strategic" choices of the path to follow upfront in other ways (how is somewhat less clear), relying on economics to help us figure out how to get there at least cost.[43]

Where does all this leave us? The social cost of carbon used by the federal government is important but may be underestimating the risks – as Stern notes, the current cost estimates should be seen "more like extreme cases; 'what might happen if we are very lucky.'"[44] And in the climate context, Canada is a high risk. It is not taking adequate steps to reduce the risk of rising temperatures. Even if we just care about economic arguments based on narrowly defined self-interest, Canada is failing.

The federal government's recent proposal to sharply increase the carbon price moves it in a better direction. According to its calculations, the increased carbon price (combined with the other policies) will allow Canada to move closer to meeting its strengthened Paris commitments. The concern, though, as we have discussed, is that its Paris commitments are seen as still too weak.

In addition to government underestimating the benefits of action by using a low social cost of carbon, many policies that do seem to be bringing about reductions are offset by other choices. BC, for example, has been justifiably praised as a leader in carbon pricing. Its GHG emissions in 2017 were only 2% lower than 2007 and were higher than in 2016, but the economy grew.[45] Emissions intensity (emissions relative to GDP) declined by about 2.5% from 2016 and almost 20% from 2007 – so the tax may just not have been enough to offset growth. At the same time, however, BC approved some large LNG projects that undercut the emissions reductions.[46]

Backsliding. While there have been moves towards stronger climate policies, some provinces and political parties have rejected this approach. In Ontario, the cap-and-trade program only ran for two years before the most recent Conservative government killed it and began fighting the federal backstop carbon price. Getting rid of the cap-and-trade system may have added almost 50 Mt of emissions, which is the same as adding thirty coal-fired plants.[47] In its place, the Ontario government put forward its own "Made in Ontario" plan,[48] which by one estimate would cost about 60% more in 2022 and about 50% more in 2030 for the same reductions.[49] Similarly, Alberta's current Conservative government is moving away from the climate policies of the prior NDP government. As economist Trevor Tombe notes, it has not killed the carbon price but did reduce its scope and level.[50] The opposition to carbon pricing in general was undercut by the Supreme Court's 2021 decision upholding the federal government's ability to set a minimum national carbon price.[51]

The result is we have a mixed bag in terms of climate policy. We use a wide range of policies but not at particularly stringent levels. Furthermore, some policies have been weakened or abandoned over time. It is not that we do not know what are the effective tools; it is that we are not using them enough. This is a familiar story. As environmental lawyer David Boyd wrote almost twenty years ago about environmental law generally, "The problem is not the lack of known legislative solutions but rather the failure to enact, implement and enforce environmental law and policies."[52] The federal government is now proposing even stronger climate action. One thing we will need to consider is how governments can credibly make commitments that everyone – businesses, individuals, other governments – can rely on. We will come back to that in the chapters on politics, but first we need to talk about why this is happening.

The Lure of Discretion

The problem is not a lack of power; Canadian legislation gives regulators the power to do any of the things we have talked about in this chapter. Instead, the concern lies in the way environmental law is structured. Canadian legislatures enact very broad environmental laws that grant discretion to someone – the Cabinet, a minister, sometimes an "independent" board – to tailor the law to particular situations. Sossin and Collins call discretion "one of the most important and least scrutinized areas of environmental law."[53] Canadian legislatures make what look to be strong laws – prohibiting pollution, allowing government officials to order pollution to be cleaned up, protecting endangered species. However, the laws include multiple off-ramps in the form of outright exceptions for certain industries as well as room for government officials to make deals in how they structure regulations or approvals or orders.

Canada has, for example, strong prohibitions on water pollution, but at both the federal and provincial levels governments can get around these prohibitions through granting approvals to companies or setting weak standards through regulations. Discretion underlies "which regulations will accompany environmental legislation, who will be appointed to environmental boards and regulatory agencies, which development proposals to approve, the nature and scope of mitigation requirements to put in place on those developments that are approved, whose voices will have input in the decision making, how statutory criteria will be interpreted, and more."[54]

The same is true in the climate change context. Most Canadian climate laws give considerable discretion to some government actor to decide on the details. Many carbon price systems are set up so that the devil is in the details and the details are subject to discretion. What is the baseline level of emissions a company must exceed before it is subject to the carbon price? It is set in the regulations made by a minister. How do we know a new oil sands mine is justified despite its greenhouse gas emissions? Cabinet has discretion to decide.

In one sense we want such discretion – we want those with expertise or information to make decisions about whether a mine or pipeline should be built. We want experts to think about how to set up a carbon price system. Legal scholar Jocelyn Stacey sees environmental law as raising the same challenges as emergencies – there is an uncertainty about what to do in very complex situations that seems to require giving someone the discretion to deal with the issue quickly and flexibly.[55] The difficulty comes when the discretion is not used to base decisions on expertise but on some other, private interest or on a mistake.

One way to think of this is as a principal-agent problem.[56] A principal-agent problem arises when someone delegates a power to act in their interests. You delegate selling your home to a real estate agent because she knows the market. You hope she maximizes the price, but maybe she just wants to sell your house quickly, so she sets a lower price or does not pay attention to selling it as she has other houses she can make more money from. You rely on her expertise but have difficulty figuring out if she is using it in your favour. The same is true when you go to a doctor or plumber or lawyer.

Canadian elected representatives enact broad laws that delegate the power to fill in the details to others with greater expertise or experience. Elected officials (in theory) hope this power is applied using that expertise. However, since legislators do not have the expertise, they have difficulty determining how it is being used and so have trouble monitoring and maintaining control of the discretion. Of course, it is slightly more complicated, as the legislature delegates to Cabinet – which is made up of elected officials. But the basic idea holds: elected representatives delegate power to somebody to make important policy choices and have difficulty controlling its use.

This difficulty with discretion arises across all areas of government policy from immigration to health care. It is problematic in the environmental area, as discretion provides an opening for industry to play a strong role in environmental law and policy. Environmental law sets up the classic collective action problem – there are heavy costs imposed on a few (industry), which gives them a large incentive to try to reduce or stop regulation. On the other hand, the benefits of taking action, though larger than the costs, are spread across a large number who are harder to organize to influence law and policy.[57]

This result is sometimes provocatively termed "regulatory capture" because industry "captures" the regulators in a way that influences policy. Industry can threaten to shut down or move to another country, can use its resources and greater information to influence regulators and, in its most extreme, essentially bribe legislators or regulators with the promise of funds or future jobs. As Boyd has noted, the consequence is that industries' private interests "prevail over the broader public interest that the government is supposed to defend."[58] Regulatory capture was a particular concern in the past when the dominant approach to making environmental law was "bipartite bargaining" – negotiations between government and industry over regulations based on the assumption that the government could be trusted to protect the public interest.[59] Part of the story of environmental law is the slow move to broaden out the participants in the policy making process, thereby hopefully reducing some of the power of industry.

Part of the issue we are wrestling with, then, is institutional – how do we get the benefits of discretion without the risk of inappropriately weak climate policies? It is about mediating between strong interests in the face of great uncertainty. We will talk more about this as we go forward. First, though, we need to understand more about the impacts of climate action on industry in Canada and how we have reacted.

One final note before we do so: this chapter has focused on one path forward – reducing emissions. We have not talked about three other paths – removing carbon from the atmosphere, geoengineering (such as making the earth more reflective in some way), or adapting to the coming effects of climate change. Canada is, and will have to continue, putting money and thought into the third. Effects from climate change are here and will continue even if Canada reduces its emissions, so we need to be as prepared for them as possible. The Canadian Institute for Climate Choices has made a strong call for significant increases in public spending for adaptation.[60] It is not the focus of this book, but it is a wise and necessary piece of the puzzle. The first, carbon removal, is also important and will become increasingly so if Canada is to reach net-zero by 2050. We can use incentives to support thinking about ways to decarbonize the atmosphere, and there is interesting work being done in this area. I am not going to talk about geoengineering except to say that (a) it is a potential option, though one with considerable uncertainty around the unintended consequences[61] and the appropriate regulatory framework; and (b) some thoughtful people argue it may be all we have left, and I dearly hope they are wrong.[62]

chapter four

Discretion II: Helping Everyone Helps No One

As is often the case when strong interests conflict, those in favour of strong action on climate change and those against seem at times to be talking past each other. Some people see climate change as an emergency that demands action regardless of the costs. Worrying about the economy is just throwing more wood into your house as it is burning down. For others, it makes no sense to take strong action when it destroys the ability to provide for those who are currently living. You may save the house but you have had to sacrifice your family to do so.

At the core of this debate is the concern that climate action will hold back economic growth and reduce jobs. This concern is not new: each time governments have sought to impose new environmental regulations alarms have been sounded that the costs are too high and will destroy some industry, if not the economy as a whole. The debate tended to be that we can protect the environment or the economy but not both.

We discussed how Canadian environmental law has come in waves, with periods of growing public concern over toxic substances or water pollution or some other issue eventually dying out as the public becomes focused on economic issues.[1] This has been called the Iron Law – when the economy is pitted against the environment, the economy wins.[2] The pull against action is even greater in the case of climate change, partly because the costs are potentially so high and are spread across all parts of the economy. Not surprisingly, this pull is particularly strong in Canada, which has the third-largest proven crude oil reserves in the world and whose oil and gas extraction industry contributes almost 8% to GDP (Canada's economic output) while energy makes up about 23% of exports.[3]

Lately, talk has turned towards the idea that we can have it all – we can both reduce emissions and grow the economy. The federal government, as we have seen, called its prior plan *A Healthy Environment and a Healthy Economy*.[4] This approach is forward looking in the sense that it seeks to foster innovation and to grow the economy on new jobs in new industries. It is hard to argue with a quest for new, higher-paying, stable jobs in an economy that uses radically less carbon.

But then came COVID-19. The idea that you did not need to make trade-offs was already controversial, but the shutdown to deal with the pandemic highlighted even more clearly the need to think carefully about the connection between climate action and the economy (including possibly the opportunity to transition to a "greener" economy). The underlying structure of the problem, though, did not change. We need to figure out who is actually being swamped by the costs of climate action and what we should do to help them. The difficulty, as we will see, is that everybody is seeking and being given shelter. The election of a new administration in Washington that takes climate change more seriously mutes some of the competitiveness concerns, but uncertainty remains.

The Unlucky Few

Strong climate action will certainly impose costs. In part that is the point – to make costly that which is currently free. The concern is that in order to make a difference in terms of climate change, the costs imposed will have to be so large that they will harm the competitiveness of industries. What happens if you add these costs into the existing Canadian economy? Answering this question requires a whole range of assumptions about how companies will react, how much we believe people care about the future, how other countries will react, and the like. Economists estimate these effects by building models of the economy or by extrapolating from past changes.[5] Both approaches tell a consistent story: the overall costs will not be large but some industries will be hurt more than others.

The Overall Costs Will Not Be Large. Most studies have not found significant overall impacts on the economy from the use of carbon pricing.[6] Putting a price on carbon to hit Canada's former 2020 goal of 20% below 2006 levels would reduce GDP by 1.5 to 3% relative to no price.[7] Estimates for meeting Canada's 2030 Paris target are more varied but range from no reductions to a 3% drop in GDP.[8] Environment and Climate Change Canada estimated that the federal carbon pricing plan under the Pan-Canadian Framework (which would not meet the former Canadian Paris target) would lead to a 0.1% decrease in annual GDP between 2018 and 2022.[9] Even for its more ambitious recent *Healthy Environment, Healthy Economy* plan, the federal government projected only a small reduction in real GDP growth of about 0.05%, which it argued likely overestimated the negative impact.[10] Studies of the BC carbon tax have similarly shown either a small net increase in employment, no significant effect at all, or only a small increase in unemployment overall but a larger increase for less-educated workers in energy-intensive industries.[11]

Another way of looking at the national impact is to think about what reducing emissions would mean to the GDP per capita in Canada.[12] The federal Parliamentary Budget Officer (PBO) examined the effect of using a $100 per tonne carbon price to meet the 30% Paris reduction in emissions. Figure 4.1 shows (real) GDP per capita increases even with the carbon price. Without the carbon tax to meet the Paris target, the PBO estimates that GDP per capita would be $61,800 per person (approximately 11.5% higher than 2014). With the carbon price, income per capita would be between $59,900 and $61,200 – lower, but both higher than the 2014 level of $55,500.[13]

Overall, then, there is an impact, but it is not enormous. Economic growth slows but continues.[14] This story is in contrast to the one often told. One important caveat is that this work was all done

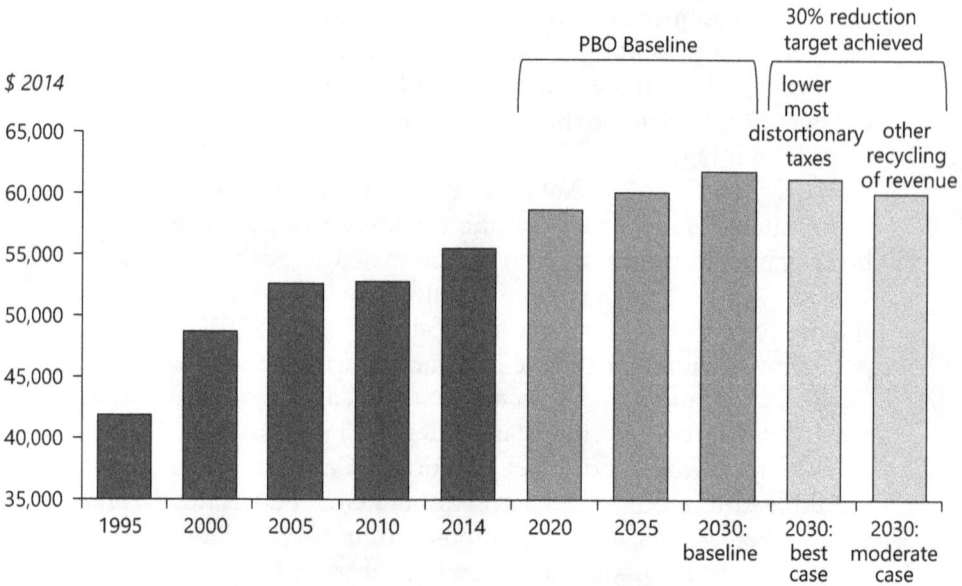

Figure 4.1. Projected GDP per capita. Revenue recycling means putting the money from any climate policy such as carbon tax back into the economy. "Lower most distortionary taxes" refers to using the revenue to reduce taxes such as on income.

Source: Office of the Parliamentary Budget Officer, *Canada's Greenhouse Gas Emissions: Developments, Prospects and Reductions* (Ottawa: PBO, 2016), fig. 5–1.

on the pre-pandemic economy. However, some early estimates of channelling recovery action into promoting a green economy suggest that such a response to the pandemic could lead to improved economic outcomes.[15]

These studies were only looking at the narrowly defined economic impact of climate action. We will have to discuss how these impacts relate to how Canadians actually live their lives and the distribution of these impacts. Moreover, to fairly assess climate action we need to take into account the costs of not acting, since climate change itself will impact the economy. These costs are trickier, as changes are difficult to pin directly on climate change, and many of them will occur farther in the future than the costs of climate action. Economist Sir Nicholas Stern, in his influential report on climate change, estimated that not taking action on climate change would cost the global economy 5% of GDP per year.[16] He noted that "the message that the costs of action are much lower than the costs of inaction is fundamental."[17] The Canadian Institute for Climate Choices recently estimated that "in the last decade, disaster costs have climbed to between five and six per cent of annual GDP growth," much of it possibly attributable to climate change, and this does not even account for the broader, more hidden costs such as to health and peoples' ways of life.[18] A 2021 federal government report noted that "while climate change will present some benefits for Canada, the associated economic impacts are overwhelmingly negative."[19] Not acting is not free.

Emissions-Intensive, Trade-Exposed Industries Will Feel It More. While the economic impact of strong climate action may not be large overall, some will feel it more strongly than others. There are two important dimensions to consider when assessing the costs of climate action to an industry. First, how much GHGs does that industry emit for each unit of output? Carbon emissions can come from the energy used or directly from the production process. Overall, Canadian manufacturing firms tend to be less carbon intensive than firms in other countries, but its energy-intensive industries (mining, pulp and paper, and non-ferrous metals) tend to be more carbon intensive.[20] The Canadian oil and gas sector, for example, has received unwanted attention for its emissions intensity. In part because so much energy is required to extract and upgrade a barrel of oil from the oil sands, it emits more GHG per barrel than average, and in fact on this measure has the fourth-highest GHG intensity.[21] The industry states that it has reduced its emissions by 30% in the past twenty years, but they remain high overall.[22]

Direct and indirect emission of GHGs is at least weakly tied to the use of energy. On the basis of how much energy is used to produce one dollar's worth of output, the industry that uses the most energy is pulp mills. The cement industry and iron and steel are also high on the list.[23] How electricity is generated in a province influences the emissions intensity of its industries.[24] Ontario, for example, benefits from having a very low-carbon electricity system, as it relies heavily on hydropower and nuclear, having shut down its coal-fired electricity-generating plants in 2014. Alberta and Nova Scotia, on the other hand, still produce some electricity with coal.

The second dimension in assessing the costs of climate action to an industry is the impact of trade on the industry: how exposed is the industry to competition from other countries, and in particular countries that have weaker climate policies? Trade exposure is particularly important to Canada as it is a "small open economy" – that is, small in economic power and highly dependent on trade. Trade constitutes about two-thirds of Canadian GDP, as opposed to about a third for the United States and Japan.[25]

Competition with other countries raises two related fears. First, Canadian industries that compete with companies in countries that do not have as stringent a climate policy will lose out. They will face costs of reducing emissions that their competitors do not. Second, any decline in production that may result from the costs of strong climate policy could be replaced by production in countries that have weaker climate policy – that is, there may be "leakage" of GHG. Canadian firms may move production to those countries, or foreign companies may be able to out-compete Canadian firms for business because they face lower climate-related costs. If so, the overall level of emissions may not go down, and may even go up.

Climate change will, then, hurt industries that are often called "emissions intensive, trade exposed."[26] The range of total economic activity that is negatively affected may not be large. The Ecofiscal Commission found that only about 5% of the Canadian economy is "more exposed," defined as sectors with carbon costs of greater than 5% of GDP (that is, the cost of meeting a particular carbon price amounts to more than 5% of the sector's output) and trade exposure of more than 15%.[27]

These impacts are not evenly spread across industries or across provinces. Figure 4.2 looks at Alberta and Ontario.[28] Industries are more emissions intensive and trade exposed as you move up and to the right. There are some similarities between the provinces in

Competitiveness Pressures by Sector in Alberta

Carbon Costs per Sector GDP at $30/tonne CO₂e (%)

100
75
50
25
10
1

LEGEND
75%
50%
25%

share of GDP   share of GHG

Refining
Other resources
Petrochemicals
Oil sands
Other metals
Coal
Cement
Fertilizer
Bitumen upgrading
Basic chemicals
Natural gas
Steel
Conventional oil
Other manufacturing
Paper
Mining
Lime

Services, government, transport & others

Trade Exposure (%)
0   25   50   75   100

The centre of each sector's bubble reflects that sector's trade exposure (horizontal axis) and its carbon costs (vertical axis; log scale). The size of each bubble reflects the sector's share of provincial GDP (grey) and share of provincial GHG emissions (black).

Competitiveness Pressures by Sector in Ontario

Carbon Costs per Sector GDP at $30/tonne CO₂e (%)

100
75
50
25
10
1

LEGEND
75%
50%
25%

share of GDP   share of GHG

Cement
Fertilizer
Refining
Petrochemicals
Steel
Other resources
Basic chemicals
Other metals
Paper
Mining
Conventional oil
Other manufacturing

Services, government, transport & others

Trade Exposure (%)
0   25   50   75   100

The centre of each sector's bubble reflects that sector's trade exposure (horizontal axis) and its carbon costs (vertical axis; log scale). The size of each bubble reflects the sector's share of provincial GDP (grey) and provincial GHG emissions (black).

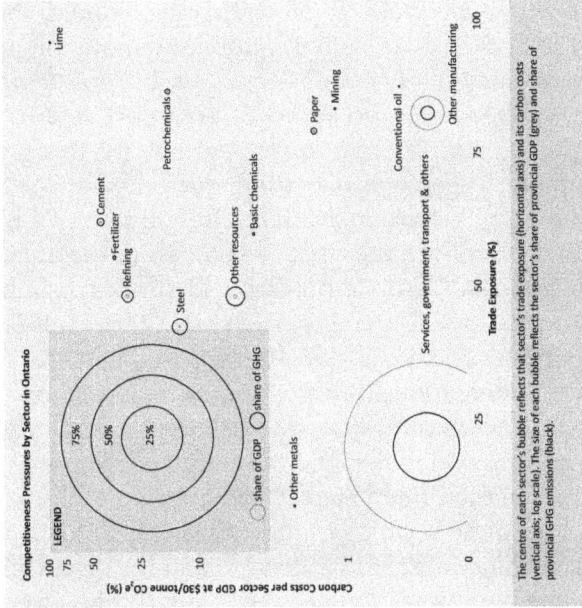

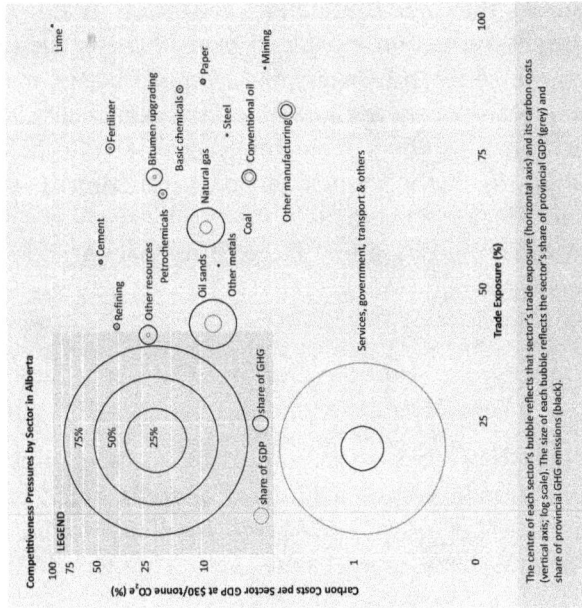

Figure 4.2. Competitiveness by sector in Alberta and Ontario.

Source: Beale et al., *Provincial Carbon Pricing and Competitiveness Pressures: Guidelines for Business and Policymakers* (Montreal: Canada's Ecofiscal Commision, 2015).

terms of which sectors lie in the upper right: lime, cement, chemicals, and fossil fuels. The most seriously impacted industries seem clear, as are those that would not be affected, the big one being the catch-all category of services, government, transport, and others. Both trade exposure and the carbon cost are low for this group, and it makes up the "vast majority of economic activity."[29]

The set of industries that are particularly harmed by climate policy is consistent across studies that use models of the economy to estimate the effects of climate policy[30] and studies examining the impacts of actual policy changes, such as BC's carbon tax or shifts in energy prices over time.[31] The models find that the Canadian sectors most likely to feel the effects of increased costs from climate policy include iron and steel, oil extraction, cement, chemicals, agriculture, pulp and paper, and coal mining.[32] These results tie in with studies in the United States[33] and elsewhere.[34]

How much will these sectors be hurt in Canada? This is a bit harder to tease out, and most studies tend to pair the discussion of impacts with possible forms of aid to the industry. However, the effect could be significant. Carbone and Rivers found some agreement across studies that a 20% unilateral reduction in emissions leads to about a 5% reduction in output from these industries.[35] On the other hand, some industries may even do better, in part because of their low use of energy as well as factors such as a lower wage rate and dollar.[36] Others have found greater effects with different assumptions in the model.[37] In part, differences across sectors will reflect factors such as whether an industry is highly competitive or dominated by a few large firms that can pass the costs onto others.[38]

What Can We Do?

If we want to help those industries most acutely affected by climate action, governments have a number of tools, often called "transitions policies." Legal scholar Michael Trebilcock called such an approach "dealing with losers," as we are trying to help those who lose from change.[39] Governments may:[40]

- *Pay the losers*: Governments could directly pay companies the costs of acting, through subsidies or tax exemptions. The payment could not only cover the costs of complying with some new requirement but help to make up for profits lost because a company's operations are more costly. An example is "revenue recycling," where the revenue from a tax on carbon emissions is given back to industries. More broadly, we could think of government paying companies to transition, such as by diversifying into other types of energy production.
- *Grant them immunity*: Government could exempt those most likely to be affected. Climate action may be required only for certain industries or for factories built after a certain date, for example. The exemption could be tied to a subsidy whereby industries receive free allocations of pollution permits under a cap-and-trade scheme – the companies avoid directly paying for pollution (since they have free permits covering their emissions).[41]
- *Delay the pain*: Relatedly, government could delay the implementation of the climate action for certain industries to give them more time to adjust. For example, in 2017 the Trudeau government proposed regulations to reduce emissions of methane, a strong GHG, from the oil and gas sector. However, when the United States revisited similar proposed limits, the Trudeau government pushed back implementation by three years (from 2020 to 2023).[42]
- *Level the playing field*: If the concern is competition from industries in countries with weaker climate rules, the federal government could impose a tax at the border on imports equal to the costs that Canadian companies have to pay. For exports, the government could give subsidies to cover the cost disadvantage for Canadian exporters. In this way, these "border tax adjustments" (or, as they are sometimes called, "border carbon adjustments") can in theory eliminate the competitive effects of different countries having different climate policies.

Which should we choose? One thing to keep in mind is why we care. We may care about the broader social benefits of keeping these industries competitive – because these industries employ many people; because their profits go to shareholders, including those who have pensions; and because they pay taxes, which help fund programs we care about such as health care, welfare, and education.

We may also feel that the only way to get industries to support serious action on climate change is to ensure they do not feel they are bearing a heavier cost than others.[43] There may be a limit on how much you can influence certain companies. The oil and gas industry may need some form of payments or protections to reduce its opposition to climate action, but given that such action threatens the existence of the industry, the cost of industry's support for strong climate action may be extremely high. Further, we may wish to help industries with their competitiveness costs if, as we discussed, we feel that they may leave Canada for a country that has even weaker GHG restrictions.

Of course, there are concerns about helping with these transitions. They involve accepting short-term benefits (generally jobs) in return for longer-term costs, as they tend to make climate action less effective. Most obviously, while helping these industries may reduce leakage, it can also mean that we do not reduce Canadian emissions by as much. The federal government's delay in imposing methane regulations on the oil and gas industry in 2017 may have helped with the industry's competitiveness concerns, but according to one estimate it will mean 55 million tonnes more methane emissions.[44] Industries may not buy new equipment where the new equipment falls under more stringent requirements or may delay early action if they hope to get payments later on from government.[45] Moreover, some of these transition policies, such as free allocation of emissions permits, mean that the government will have less revenue to spend on other things, including emissions reductions; support for new technologies; health care, education, or social programs; or reductions in other non-environmental taxes (such as income or sales taxes).

In addition to potentially leading to less effective climate action, transition policies may be unfair. Companies make decisions such as where to invest and how much pollution prevention equipment to install based on a calculated risk that the underlying policy framework will not change. Some feel these companies are rational, sophisticated actors and should be required to live with their decisions. It is not clear why companies should be compensated when their choices (such as to invest less in pollution control) do not pay off, particularly when it means that someone else will have to bear the costs.[46] This sense of unfairness may be reduced where the changes are unanticipated,[47] but that seems less true for climate change in recent decades.

This unfairness may be compounded if only certain "politically influential losers" benefit from transition policies.[48] Some companies and industries have greater resources and political power to lobby for weaker limits, exemptions, and compensation. Those with fewer resources and less power are less likely to get relief. We then come back to our story of "regulatory capture." This unfairness does not only exist across companies and industries. Vulnerable individuals, including workers in industries that are not provided with relief, may not receive relief, nor may Indigenous groups or less powerful consumers. [49] And their losses may go beyond the economic: climate change could lead to loss of species or alterations in traditional lands.

We need to consider how best to use the "costs" of our transition programs. These costs include direct payments (subsidies) or foregone payments (such as giving away allowances for free), but also the increased GHG emissions that will have to be offset by someone else (or if not, the costs from the added risks from climate change). We should try to get the most bang for the buck for those costs. If there are industries that may never be able to fit within a carbon-free economy, we need to consider how much the short-term jobs and revenue are worth. The alternative to supporting these industries to continue along the same path is to spend that money or emissions in helping other industries become less carbon intensive.[50]

That is not to say we do not support those industries that are not likely to exist in a carbon-free economy. We may want to provide time for resources and workers to move away from them. But we need a conscious focus on where we are using our resources.

Who Are the Lucky Losers?

If we are worried about the competitive impacts of strong climate action, we would want the relief to be:

- focused on those who actually face competitive pressures or need relief from costs[51];
- temporary to reflect the fact that we are transitioning;[52] and
- structured in a way that minimizes harm to the environment.

Unfortunately, we fail on all these measures:

Help Is Not Targeted. Relief has not been given just to those who face competitive pressures. The government may feel it needs to give broader relief in order to get buy-in for stronger climate action, but it is not clear how much of that support is actually arising. Canadian climate policy is not strong and yet governments are giving relief which undermines the relatively weak policies that do exist.

The distribution of relief is hard to tease out. In one sense, everyone gets "relief" from regulatory costs because the targets have been set so low. The Paris commitment, for example, could be seen as transitional relief to a longer-run decarbonization of the economy that Canada states it seeks to achieve by 2050.[53] Even if we felt the targets were adequate, Canadian governments have spent a long time doing nothing – sparing everyone the costs of acting. The federal government made plans but the plans were weak and not realized. As political scientist Kathryn Harrison has noted, "Confronted by inattentive voters, opposition from greenhouse gas intensive industries, and provincial governments defending local industries, Canadian governments have relied almost exclusively on politically appealing but ineffectual voluntary programs and subsidies."[54]

Even when Canadian governments did decide to act, they limited the effectiveness in two main ways. First, they used caps to limit the harm to industry. Prime Minister Chrétien, for example, explicitly promised industry that the climate change costs would not exceed $15/tonne. This cap did not relate to industries that

were particularly vulnerable to competitive pressures but covered everyone. The Harper government effectively continued this
promise. The BC government has recently revisited this price cap
idea by holding the tax for liquefied natural gas at $30/tonne.[55]

Governments have also paradoxically used no caps as a method
to provide relief. Under the federal government pricing policy, for
example, companies pay a price if their emissions per unit of output are above a certain threshold. However, emissions intensity can
go down but emissions still rise if output grows faster. In fact, that
has been a big part of the climate change story in Canada –national
emissions intensity has gone down but that has not stopped the
rise in emissions. If the Liberal government actually puts a cap on
emissions for oil and gas (as opposed to limiting intensity) as they
have promised, it would be a significant new approach.

Second, both federal and provincial governments have exempted
various industries from requirements to take climate action but the
exemptions may not be tied to actual competitiveness concerns.
For example, the BC government initially had its carbon tax cover
almost all industries. However, the BC agriculture industry asked
and in 2012 was granted relief from the BC carbon tax even though
there is no evidence it faced harm.[56] Ontario's now-cancelled cap-
and-trade scheme was another case in point. While it had covered
82% of emissions,[57] at least in the initial period most covered parties received free allowances so in effect were given a subsidy as
they could sell these allowances.[58] Yet by one measure less than
2% of Ontario's GDP is actually exposed to risk from carbon pricing.[59] The government explicitly noted that in the first period free
allocation would not be connected to the risk of competition and
leakage.[60]

The current federal backstop carbon pricing plan is a bit different. The plan puts in place a system for large emitters (over 50
tonnes per year) that covers all industries. The basic idea is that
it sets a threshold for action that is tied to the national average
emissions per unit of output for each industry. Companies with
lower emissions than the threshold receive allowances they can
sell. Those with higher emissions have to pay the price. The threshold was initially set at 70% of the national average. After lobbying

from industry, this threshold was raised to 80% for some industries and 90% for others.

In part the extra relief was for industries at risk. The 90% threshold applies to the cement, iron and steel manufacturing, lime and nitrogen fertilizer sectors[61], all of which plausibly are at risk from climate measures in the sense of being emissions intensive and trade exposed. There seems no principled reason to shift up the threshold for all other industries, whether or not they face actual competitiveness pressures.

Federal and provincial governments then seem cognizant of competitiveness complaints. However, economists Sarah Dobson and Jennifer Winter note "appropriate targeting of EITE support policies has largely been overlooked in most jurisdictions."[62] We know who the industries at risk are yet governments give relief to everyone.

Relief Does Not Appear Temporary. It is also not clear that the relief that has been given is temporary. We may need relief policies to allow the Canadian economy to adapt in the face of lack of action by other countries.[63] This relief should be reduced as the adjustment occurs. Some of the delayed implementation, such as for the methane regulations, seems aimed at allowing industry to adjust to the increased costs over time. For other policies, there is no end in sight.

There is, for example, no sense in how long transition relief will exist under the federal backstop; the federal government says it will reduce these thresholds over time but not when.[64] The initial carbon price is to be in place until 2022. However, the relief (the threshold for when the price kicks in) is in guidelines as opposed to legislation so it could either be increased or decreased easily over time. In recently proposing to increase the carbon price by $15 per year until 2030, the federal government appears to be keeping the same underlying scheme. Moreover, the federal government has certified that most provincial pricing plans are equivalent to the federal plan, yet each has its own system for exempting industry and it is not clear if the federal government will require any form of reduced exemptions.

Even where there are phase-out plans, they may not be tied to actual changes in competitiveness conditions. Under the former

Ontario cap and trade program, for example, the free allocation of allowances was intended to be time limited and to decline slowly. Ontario planned to reduce free allowances by 4.57% per year for most emitters – meaning they would have had to purchase more allowances or reduce emissions. It is not clear whether this rate of reduction, though, was tied to competition concerns.[65]

Relief Does Not Minimize Harm. The last condition – that the relief should be given in the least harmful way for the environment – is even harder to measure. We can provide some relief for those facing the costs of climate action without completely undermining the goals. We can order relief policies, in a very rough way, from better to worse: [66]

- *Border tax adjustments (BTAs)*: As we discussed, border tax adjustments target trade flows directly, taxing imports at a level that matches the cost imposed on Canadian firms by climate policy or paying Canadian exporters to cover their costs.[67] If done properly, they can mean Canadian producers are not disadvantaged relative to countries that do not take climate change as seriously. However, these border measures can be difficult to design to ensure they comply with our commitments under international trade agreements to ensure they are not being used to protect Canadian domestic industries (as opposed to just offset the costs of weaker foreign carbon policies).[68] Fortunately, the case for border measures is clearer when they are used to offset carbon prices, which the federal government is leaning on, as opposed to the costs of regulations. In terms of effectiveness, one study found that if Canada on its own implemented a BTA to offset the effects of carbon pricing, it would produce the largest loss of GDP when compared to other policies.[69] A solution that overcomes some of the concerns about BTAs is to band together with other countries willing to take climate action and jointly impose these BTAs on countries that are not taking action.[70] Such "climate clubs" might mitigate competitiveness effects for those within the club and give greater incentives for action for those on the outside. These border measures, though, decrease trade and so can raise fairness concerns about who is inside and who is outside the club.

- *Output-based systems*: The federal government is using this output-based method in its backstop legislation, tying reductions to emissions intensity. Economists prefer this method of relieving competitiveness concerns as it provides an incentive to reduce emissions without constraining output. [71] Because there is no cap on emissions, the price might need to be higher to ensure there are emissions reductions.[72] Further, it may not work for some sectors such as electricity and petroleum as the implicit subsidy may keep the price of electricity or petroleum low, reducing the incentive to reduce consumption and possibly resulting in higher emissions.[73] The costs of designing and implementing an effective system may offset what could be minimal gains.[74] Accordingly, it may be one of the best in terms of environmental impacts but only if you can get a high enough price to make it worthwhile.
- *Revenue recycling*: This tool involves payments to affected industries to overcome the competitiveness risks. Lump-sum payments should not distort the firms' decisions to, for example, increase or decrease production. There are different versions of this tool. Some involve not payments but cuts to various taxes. The evidence is mixed as to whether policies such as reducing capital taxes aid in combatting leakage.[75]
- *Exemptions*: These are the worst form of relief for competitiveness concerns. Exemptions in the form of grandfathering (exempting existing firms) can deter the birth of new firms not covered by the exemption.[76] More importantly, they can exempt cost-effective reductions and so force others to take more costly actions.[77] The broader the exemptions, the harder and more costly it is to meet any emissions targets.

Unfortunately, the federal and provincial governments have relied most heavily on the worst forms of relief – exemptions and subsidies – and have been reticent to use potentially more effective (though complex) forms such as border tax adjustments. In addition to the extension of weaker requirements to all industries we have already discussed, governments have responded to complaints from industry by delaying the introduction of climate measures such as restrictions on methane. Sometimes the delay rationally provides time to adapt to costs; sometimes it amounts to an inefficient payment to industry.

This pattern of exemptions or delays continues. The federal government's planned Clean Fuel Standard was promised, touted as a major tool but delayed until after the election and then again because of COVID.[78] A number of provinces delayed planned increases in the carbon price to provide relief during the pandemic. While help to industry was clearly needed because of the economic effects of the pandemic, the question is whether this is the best form of relief given its negative longer-term effects on the environment.

If governments exempt some industries to help them with competitiveness concerns, Canada can only meet its emission reduction targets if governments increase the price or the restrictions on other industries. Whether or not this is useful will then depend on the size of the emissions that are exempted and how large the impact has to be on other sectors or activities to make up for this exemption.[79] The oil sands, for example, are a very large source of Canada's emissions now and projected to be so in the future. To make up for an exemption of that sector would require significant reductions by others. It would exempt potentially large cost-effective reductions and force others to take more costly ones.[80] If the exempted sectors are facing competitive pressure, there may be some offsetting gains from the industry producing more (and employing more) than it would without the exemption. If the exempted sectors are not facing competitive pressure, there are no such offsetting gains. Again, the federal promise to cap oil and gas emissions would at least ease some of the impact on other industries. However, note that the baseline for the cap seems to be at best 2021 or 2022 oil sands emissions, exempting the industry from the 2005 baseline for the country as a whole.

Even if the exemptions were tailored to competitiveness concerns, the immunity from action and the delays in implementation undermine both environmental and cost-effectiveness goals. While the federal government may be using the "best" form of output-based pricing, it undermines its environmental effectiveness if the carbon price is kept low. The proposed price increases in the new federal plan would remedy this concern somewhat.

Subsidies are another life preserver thrown to industries whether they are drowning or not. As we discussed, the federal government under both Liberal and Conservative parties has long relied on subsidies. The early climate plans in fact used payments as a central tool to try to reduce emissions. The most recent federal plans also rely heavily on subsidies, supporting green technology and decarbonization through investment and direct government purchasing. The subsidies seemed tied in least in part to some notion of growing a green economy[81] but it is not clear the decarbonization funding was tied to competitive pressure concerns. Its *Healthy Environment, Healthy Economy* plan refers to both general industrial support and "targeted support to large emitters in the oil and gas sector, cement, iron, and steel sectors."[82]

Finally, the federal government plan also states it is "exploring the potential of border carbon adjustments" with "like-minded economies."[83] While these measures have benefits in terms of addressing competitiveness concerns, it is unclear whether their use will lead governments to reduce the other relief mechanisms already provided to potentially vulnerable industries. Moreover, the concern, as with most trade measures, is that they will be used not to offset legitimate risks of leakage or environmental impact but instead to become a method by which industries can press for protection against international competition, whether they are risk because of climate policies or not.

Discretion Again

Canadian climate policy, then, again looks, at least in part, like the political economy story – the government wants to buy off most industries and therefore sets caps on costs, grants broad exemptions, and gives funds to affected industries. Some of the relief tracks actual competitiveness concerns, but more seems to be across the board for industry. And government has failed to use one of the best tools – border tax adjustments.

This approach to transition relief is unfair to those who have to make up the difference. It seems to be a windfall to more powerful, industrial interests with costs off-loaded onto less visible, more diffuse, less powerful actors. There is also, given the exemptions to carbon pricing, less revenue to aid the vulnerable in transitions. This undermines efforts to meet the targets in a cost-effective fashion, shifting the burden into the future (which is likely to be costlier) or onto those who have higher costs of emissions reductions. The result may be short-term political buy-in but potentially at a high cost. And there appears to be little added buy-in from some industries or actors.

Part of this is due to the institutional story we discussed last chapter. Politically accountable actors in general decide on the exemptions – politicians would like to get the credit for and have the benefit of doling out relief. Industry pushes hard for relief, whether warranted or not. As a result, larger longer-term benefits are traded off for smaller short-term economic gains. Again, the discretion at the core of environmental law provides a fertile ground for such myopic decisions.

There is no actual path that leads to climate action without concern for those who lose in the transition. None of the transition tools are perfect, and most of them are not as tailored as we would like – but we need to help the right losers.

chapter five

Diffusion: When Everyone's Responsible, No One's Responsible

When you read about Canada's climate change plan, you tend to hear that Canada has signed on to the Paris Agreement target and that there is a national plan in place to meet this target. The federal government and the provincial governments are trying to find solutions to reduce carbon emissions. It sounds as if Canada is like a family trying to live within a budget. The parents set the budget and everyone is trying to tighten their belts to do their part. The parents are cutting back on expensive dinners out, movies, and vacations while the kids have to live within a smaller allowance and a more limited data plan for their phones. It's hard but they are all in it together.

In reality, Canada is a much more dysfunctional family. The federal government agreed to the Paris targets and has tried to develop a plan that encourages reductions, including requiring the provinces to adopt a carbon price. Some such as BC and Ontario had taken significant steps to reduce emissions, in BC's case by imposing a carbon tax and in Ontario by shutting down coal-fired electricity generation. Many of the other provinces have taken some steps to cut back on emissions but failed at least initially to specify detailed policies.[1] Alberta and Saskatchewan are in a different situation. They have developed plans to reduce emissions but, given the nature of their resource industries, their emissions have been rising, and it would be costly to achieve emissions reductions equivalent to those of other provinces. The federal government has tried to find a consensus position with the provinces throughout this process, but agreement has been diffi-cult to find and fragile.

Differences across provinces are a major problem for coordi-nated climate change action in Canada. The federal government's solution in the past has been to let everyone do their own thing. Provinces get to set their own targets and figure out how to meet them. Moreover, the federal government at times responded to the demands of different provinces to water down regulations that hurt them in particular – such as Alberta's concerns about oil and gas regulations and Ontario's about restrictions on vehicle emissions.[2] The result was that almost no action was taken. The federal gov-ernment's Pan-Canadian Framework was an attempt to set a mini-mum set of rules. Unfortunately that small apparent consensus fell apart and, even if it had not, there would still be questions about whether the framework represents a well-thought-out, principled approach to sharing the costs of climate action across Canada. The federal government's new *2030 Emissions Reduction Plan* frame-work proposes a stronger price signal and more subsidies but only marginally adds to discussion of how the costs should be split.

This chapter looks at this difficulty in finding a national approach and the options for sharing costs across provinces (we will discuss Indigenous governance in later chapters). Before we talk about this, though, we need to get a better sense of how the costs vary.

A Different Set of Solitudes

Hugh MacLennan wrote a famous Canadian novel called *Two Soli-tudes* about the lack of communication and understanding between French and English Canada. Canadian climate policy raises a new solitude based in part on what political scientist Douglas Macdon-ald calls the "West-East divide."[3] We can see the divide in figure 5.1, which shows considerable differences in levels of GHG emissions by province over time. Emissions are a function of economic activ-ity and population; the larger provinces tend to have more of both and so higher emissions.

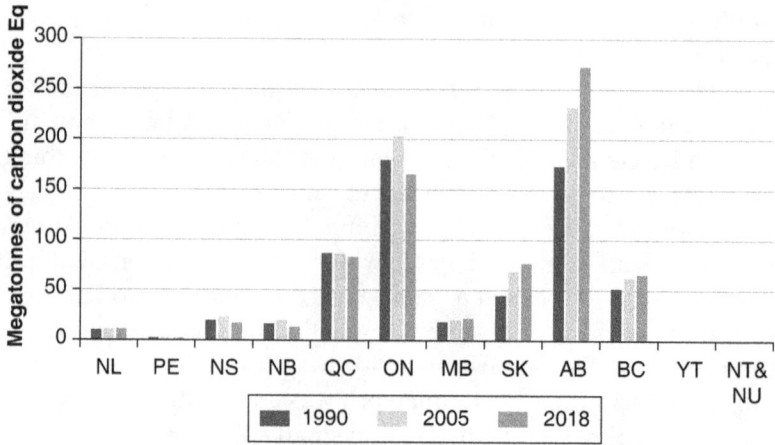

Figure 5.1. GHG emissions by province in 1990, 2005, and 2018.
Source: Environment and Climate Change Canada, *Greenhouse Gas Emissions, Canadian Environmental Sustainability Indicators*, 11, fig 7.

Not only does the difference in the size of the emissions show the divide, but so does the path of emissions over time. Alberta and Saskatchewan have significantly increased emissions since 2005, with BC, Manitoba, and Newfoundland having smaller increases. Alberta's emissions rose by about 18% between 2005 and 2018 and Saskatchewan's by 12%.[4] In contrast, both Ontario and Quebec's emissions fell. Basically, since 2005 (the baseline for the Paris Agreement) five provinces have reduced emissions or remained constant and five have increased.

Another way to see the divide is by looking at the per capita emissions shown in figure 5.2. Again, Alberta and Saskatchewan stand out: they have significantly higher per capita emissions than any other province. The rest are between 10 and 20 Mt per person each year, while Alberta and Saskatchewan are over 60 Mt. The difference comes from the nature of their economies – both Alberta and Saskatchewan have relied heavily on fossil fuel production.

These differences in emissions mean that provinces face very different pressures in trying to reduce them. As we saw in the last chapter, industries vary by their carbon intensity and their

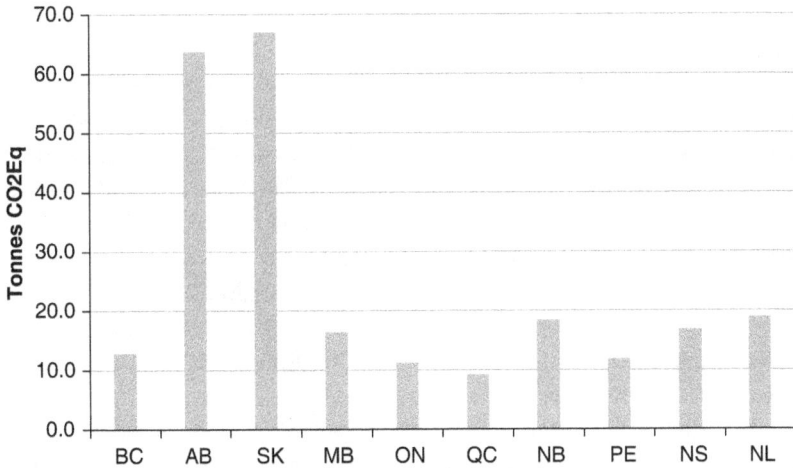

Figure 5.2. Per capita GHG emissions, by province, 2017.

Source: Environment and Climate Change Canada, *Greenhouse Gas and Air Pollutant Projections: 2018*, https://www.canada.ca/en/environment-climate-change/services /climate-change/greenhouse-gas-emissions/projections-2018.html, and Statistics Canada, "Canada at a Glance 2018: Population," https://www150.statcan.gc.ca/n1 /pub/12-581-x/2018000/pop-eng.htm?HPA=1.

vulnerability to competition from other countries. The differences in emissions across provinces are tied in large part to which industries are important in each province. Look again at the differences between Alberta and Ontario. Figure 4.2 (page 67) shows the energy intensity and trade exposure of different industries for each province, with those industries in the upper right of the figure being most vulnerable.

For both provinces, a fair share of GDP is not vulnerable – coming from sectors that include service, government, and transport. However, Alberta has a greater share of both GDP and emissions in the upper-right quadrant, where industries are both energy intensive and trade exposed. The oil sands are obviously in that area, but so are a number of other industries that are important to the Alberta economy, including natural gas, fertilizers, and petrochemicals. Ontario, on the other hand, has some similar vulnerable sectors but they are not as important economically. Oil and gas being the top source of emissions in the country has a significant impact on these differences.

Not surprisingly, emissions intensity tracks factors such as the source of economic activity and of energy. Alberta and Saskatchewan have the highest emissions intensities, reflecting their reliance on the oil and gas sector as well as, for Alberta in particular, coal-fired electricity generation. Alberta has the lion's share of the oil and gas sector in Canada, accounting for about three-quarters of Canada's oil and gas production.[5] It produced almost 70% of the country's marketable natural gas, with BC second at almost 30%.[6] At the low end of emissions intensity is Quebec with its electricity based on hydro power.[7]

Reliance on a particular resource or "staple" can be a mixed blessing. When oil prices skyrocketed between 2002 and 2008, Alberta (and Saskatchewan and Newfoundland and Labrador) saw large growth in provincial income, though other provinces were hurt by the corresponding higher exchange rate.[8] On the other hand, the recent oil price drop has hurt Alberta, with unemployment rising to 9.1% in 2016 before falling back to 6.6% in 2019. Alberta was dealt a further setback by the oil price shock in 2020 coupled with the pandemic, with unemployment in June 2020 rising to almost 16%.[9] It remained high but slightly lower at about 11% by December 2020 before falling somewhat in 2021.[10]

Figure 5.3 shows the Ecofiscal Commission's estimates of how provinces differ in levels of GDP that is exposed to the costs of climate action. Just about 4% of GDP is exposed on average across Canada; only Alberta and Saskatchewan have a share of GDP that is more exposed than that level, at about 18% each. Ontario and BC are at about 2%, while Quebec is at 1%.[11] This does not mean that 18% of the Alberta or Saskatchewan economies will disappear if strong climate action is taken, but it does point to a much higher level of economic costs of action in those provinces. Similarly, the Canadian Institute for Climate Choices found the highest percentage of workers in "transition-vulnerable sectors" in Alberta, the Northwest Territories, and Newfoundland and Labrador.[12]

All this so far speaks to what has been – what emissions actually were in each province. All provinces are taking steps to address

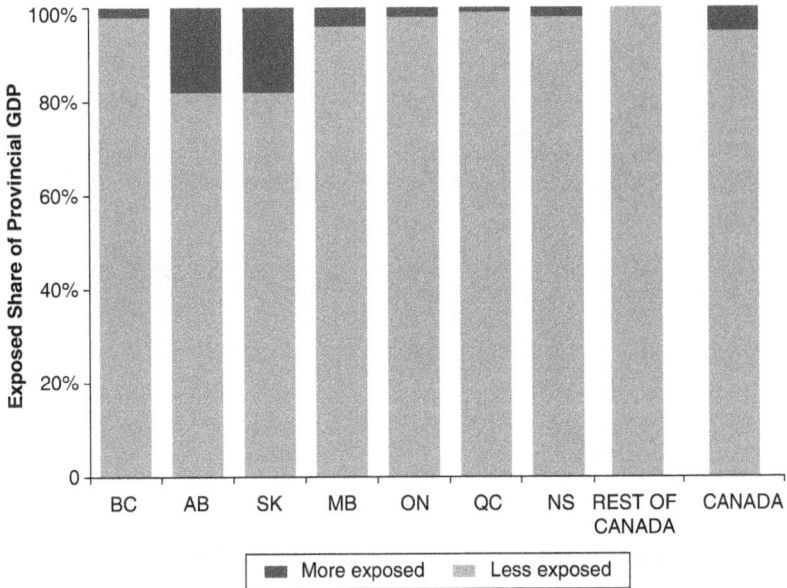

Figure 5.3. Share of GDP that is energy intensive and trade exposed (2015).
Source: Beale et al., *Provincial Carbon Pricing and Competitiveness Pressures: Guidelines for Business and Policymakers* (Montreal: Canada's Ecofiscal Commision, 2015), 13, fig. 2.

climate change. What about the future? It turns out emissions will likely grow in some provinces. Alberta, for instance, has set a cap of 100 Mt from the oil sands. However, the Pembina Institute noted in 2018 that, while the GHG emissions from the oil sands was 77 Mt, Alberta had approved projects that will add up to 131 Mt if all go ahead.[13]

As a result, Environment and Climate Change Canada predicted a very familiar future. Figure 5.4 shows the federal government's projections for each province by 2030 taking into account measures that are actually in place in September 2019. Based on measures in place, most provinces drop a little between now and 2030. However, the end result is that Canada as a whole is way off its Paris targets, as the provincial emissions add up to 673 Mt, well over the Paris target of 511 Mt.[14]

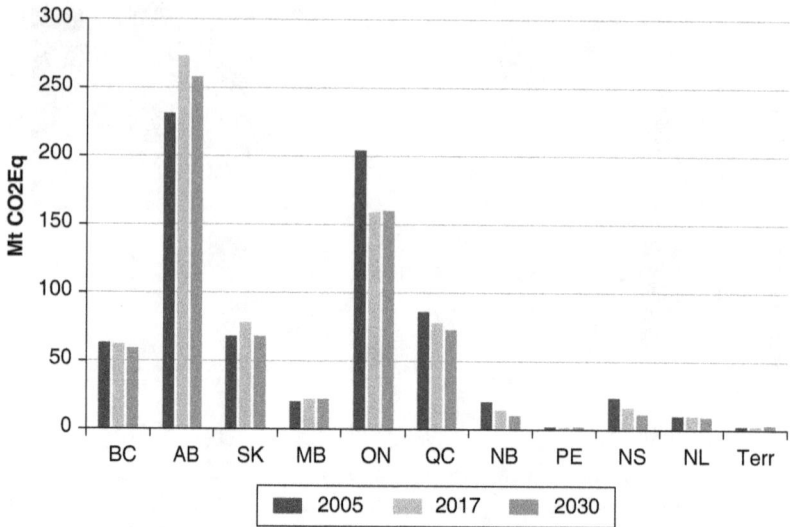

Figure 5.4. Provincial and territorial GHG emissions, 2005–2030. These projections by Environment and Climate Change Canada take into account economic factors and government policies on climate change in place as of September 2019. The projections do not include land use, land-use change, and forestry activities.

Source: Environment and Climate Change Canada, *Canada's Fourth Annual Biennial Report on Climate Change* (Gatineau: ECCC, 2019), table 5.8.

Things were a little better if you took into account measures that were announced under the Pan-Canadian Framework but not yet in place.[15] Almost all provinces will have stabilized or reduced their emissions from 2005, with one exception: Alberta continues to have higher emissions than in 2005 under the plans that have been announced. Even with these announced measures, however, Canada was still off its former Paris target of 30% below 2005 levels by 2030.

More recently, the federal government strengthened a number of policies, including the carbon price. It predicted the plan would bring Canada's emissions within the former Paris commitment. With measures in place as of September 2020, some provinces continue to have emissions in 2030 that are higher than 2005 levels, including Alberta (up 6% over 2005 levels), Manitoba (6% higher), and Newfoundland and Labrador (2% higher).[16] However, the federal government subsequently increased its targets to 40–45% below 2005 levels by 2030 and achieving net-zero by 2050. With oil and gas being the largest source of emissions in Canada, the federal government has promised to cap emissions from this sector, which will impact Alberta and Saskatchewan in particular. Beyond this potential cap, however, the federal government has not explicitly negotiated how the added emissions reductions will be spread across the provinces and, as we will discuss, at times announced reductions without apparent discussions with the provinces.

How Could We Split It Up?

Emissions, then, vary considerably across the country. The approach to reducing emissions so far has been "piecemeal," resulting in a lack of a coherent national approach to assigning emissions reductions.[17] The costs of reducing emissions could be split up according to a few different principles: efficiency, rights, and fairness.[18] These may overlap, but they are good general categories.

Efficiency: One way to think about who should reduce emissions is to ask what would allow us to meet our Paris targets at the lowest cost. If we can minimize the overall cost of action, Canada as a whole will be better off. Given what we have seen about the different industries and emissions patterns across provinces, the least-cost solution might still involve very high costs for some regions. Before we ratified Kyoto, the federal government focused on allocating costs by sector as opposed to by province. Focusing on a sector (such as the steel industry, for example) was viewed as efficient, as the sector would face the same standards across the country.[19] Another way to view an efficient allocation is to consider the emissions resulting from a uniform carbon price across the country.[20] The Pan-Canadian Framework, and the subsequent *Healthy Environment, Healthy Economy* plan, took a version of each of these approaches – they set a minimum carbon price, along with standards for some emissions such as from coal-fired electricity generation. As we will see, though, this approach leaves much of the work to the discretion of each province, which may undermine efficiency.

Rights: Alternatively, we may view the atmosphere as a shared resource and believe every Canadian has an equal right to use it. This notion of sharing emissions on a per capita basis, with larger provinces getting a correspondingly larger share, has been called "egalitarian."[21] It would not take into account the differences in provinces we talked about, such as the resources they have or the extent to which they are exposed to competition from other countries.[22] The costs of reduction do not matter on this approach; the only important fact is how many people live in a province.

A different version of a rights-based argument is more prevalent – a province's share of future emissions should depend on what it has emitted in the past. The argument here is, in essence, that a province has a right to its emissions levels and that we should use that as a baseline for doling out shares of future emissions. It ties into the fact that the Constitution gives ownership of and power over natural resources primarily to the provinces. Because based on historic emissions, this approach is also called "historic," "sovereignty," or status quo.[23] We will call it the "proportionate" approach to capture the notion that it involves all provinces reducing emissions a proportionate amount based on their prior emissions. One difficulty with this approach is that it depends heavily on what year you are using as the baseline – those who have not acted will want the current year while those who reduced emissions in the past will want the baseline set prior to whenever they made reductions.

Fairness: Fairness more explicitly considers the distribution of the costs and benefits. A "Rawlsian"-inspired approach would, for example, aim to minimize the harm (or maximize the benefit) to the least well-off.[24] It could be designed to do the least harm to the least well-off individuals or the least well-off provinces. On a provincial basis, for example, the reductions Canada has to make (or, equivalently, the emissions Canada is permitting itself) would be distributed so as to leave the poorest province at the best level possible.[25] A more easily implemented version of this notion of fairness would be to base allowable emissions on the wealth or resources of a province. Poorer provinces may be required to make fewer reductions or even be allowed to increase emissions.[26]

Finally, an alternate approach to fairness would be polluter pays. On this view, Alberta and other provinces benefited enormously from exploiting their oil and gas resources. They did not save for the future (as Norway has done with its sovereign wealth fund based on its oil revenues) and, in fact, built up liabilities – by increasing public spending without increasing other taxes and by allowing large potential future environmental concerns in the form of massive tailings ponds. We will return to this notion later. It is deeply held by some but in many ways is not helpful.

Not surprisingly, the provinces bear different costs depending on the approach chosen. Figure 5.5 provides a rough measure of the differences in emissions across the provinces under the different rules, from which a few things are apparent. First, Alberta and Saskatchewan get the greatest share of emissions under the proportionate rule. Conversely, they both would have to reduce emissions the most under the per capita rule given their large emissions and relatively small populations (This is consistent with a study estimating where the costs would fall if Canada attempted to meet its 2020 Copenhagen goals).[27] Alberta, Saskatchewan and, to a lesser extent, Newfoundland and Labrador are asked to do more under all rules except the proportionate rule, as they have higher emissions and lower-cost means for reducing emissions. Second, most of the other high-emitting provinces tend to get the lowest share of emissions under the proportionate rule but the greatest under the efficiency rule, as the efficiency rule allocates more to provinces with lower emissions intensities (emissions per dollar of GDP). In theory, these provinces have higher costs of further reducing their emissions than ones currently emitting more for each dollar of GDP. Finally, Nova Scotia and New Brunswick get the most emissions under the fairness rule, which allocates more emissions to provinces with lower GDP per capita.

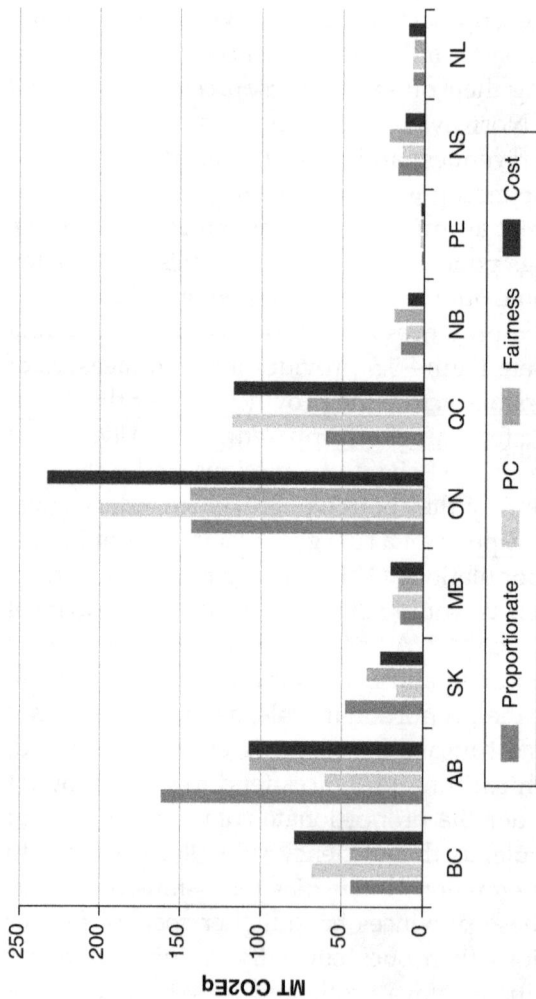

Figure 5.5. 2030 provincial emissions under various sharing rules. "Proportionate" is 30% below the 2005 emissions levels for each province (that is, each province takes on the national goal). "PC" divides the 2030 target level of emissions across all provinces on a per capita basis. "Fairness" alters the "Proportionate" level of emissions such that provinces with lower GDP per capita get a greater share of emissions (and with higher GDP per capita get a smaller share). "Cost" alters the "Proportionate" level of emissions such that provinces with higher emissions intensities (emissions/GDP) get fewer emissions (and lower emissions intensities get greater emissions).

Source: Environment and Climate Change Canada, Canada's Greenhouse Gas and Air Pollutant Projections Emissions (Gatineau: ECCC, 2018), 17, table 8; Statistics Canada, "Canada at a Glance 2018: Population," https://www150.statcan.gc.ca/n1/pub/12-581-x/2018000/pop-eng.htm (provincial population data for 2017); Statistics Canada, "GDP at Basic Prices," https://www150.statcan.gc.ca/t1/tbl1/en/tv.action?pid=3610040202 (GDP for 2018).

How Are We Splitting It Up?

It is hard to know which of these rules is being applied, in part because there has been no recent formal discussion of how reductions should be spread across provinces. As we will see in chapter 10, the federal government signs international treaties for the country, such as the Kyoto Protocol and the Paris Agreement, but does not have a clear constitutional power to implement them – the Constitution does not explicitly assign the power to make environmental laws to either the federal or provincial governments. The federal government has some specific powers such as over fisheries as well as broader powers over issues like criminal law and laws that affect the "peace, order and good government" of Canada. The provinces can make laws dealing with local matters and property and have ownership of natural resources on provincial lands, which turns out to include most aspects of sectors like oil and gas and forestry. Power is diffused across the federal and provincial (and to some extent territorial) governments.

While the federal and provincial governments did begin discussing who should bear what costs in 2002 prior to Canada's signing the Kyoto Protocol, they never reached a conclusion.[28] They drafted an agreement stating that no region would bear an "unreasonable" burden and indicating that some form of relief would be given to those most affected, but never signed the agreement and have not entered into a formal allocation framework since then. The result is that federal and provincial governments make their own climate policy, with provinces setting their own targets.

The Federal Plan

The Pan-Canadian Framework was the federal government's plan for addressing climate change.[29] Almost all of the provinces signed on to the framework, so in effect it was a "national" plan. It does not, however, specifically discuss the allocation costs across provinces, beyond vague statements about it's being a "collective plan" and a "collaborative approach" that gives the provinces "flexibility." The closest it comes is reaffirming the principle of collaboration while "recognizing the diversity of provincial and territorial economies and the need for fair and flexible approaches to ensure international competitiveness."[30] It mentions the need to "minimize competitiveness impacts and carbon leakage" and "consider and mitigate impacts on emissions-intensive trade exposed sectors."[31] The federal government's more recent *Healthy Environment, Healthy Economy* plan again has no set allocation principles, stating only that "a deep respect for, and recognition of, shared constitutional jurisdiction will continue to be an important guiding pillar in all of Canada's efforts to fight climate change."[32]

The federal plans, then, recognize that everyone has to play a part and that somehow the division of responsibility should be "fair" and "flexible." They implicitly contain some decisions about who should bear the costs of reducing emissions. For example, the common minimum price approach means that the federal government is (partially) committed to the idea of efficiency in reductions. The approach is not completely efficient, since it is not a truly common price. Provinces with their own similar pricing systems can keep those systems in place. Each includes relief for their own industries, and the federal government has been very flexible (some say too flexible)[33] in determining where the systems are similar. For those provinces without a carbon price, the federal government's backstop pricing plan includes relief for different industries. As we saw last chapter, the pricing relief seems less about helping emissions-intensive, trade-exposed industries than helping politically powerful ones. It shifts the pricing system towards the proportionate approach.

The federal government also has some existing and proposed nation-wide regulatory standards. For example, it enacted a nation-wide phase-out of coal-fired electricity generation by 2030 and limits on methane emissions, and is working on a clean fuel standard. More recently it proposed requiring 100% net-zero electricity generation by 2035. Such uniformity could be seen as a move towards one version of an efficiency approach – a whole sector faces the same requirements nationwide rather than trying to tailor them to a particular region. Again, the federal government has softened the approach to take account of the costs to particular industries or provinces – so, uniform but possibly anemic across a sector.

Finally, the federal government's recent promise of a cap on oil and gas emissions relies on a version of proportionate share. The sector (and provinces) will be permitted to maintain their emissions (which is not explicitly the case currently) and then reduce these emissions over time (at an unspecified rate). Instead of a 2005 baseline as in other sectors (or provinces), the baseline will be current levels of emissions. The result is a modified proportionate approach recognizing a form of historic right to a level of emissions.

The Provincial Plans

The federal government tries to set a common ground for some reductions but the provinces set their own emissions reductions targets and plans. Have the provinces made up the difference in terms of the targeted emissions reductions? The short answer overall is no. The provinces have set targets that will not, if added all together, reduce emissions to a level that meets our Paris commitments – and that is even before we look at whether they have plans that will actually let them meet those targets.[34] In fact, the Pembina Institute found that over 50% of Canada's emissions are not covered by a target for 2030 – and 75% for 2050.[35]

One way to see the allocation of emissions is to examine Environment and Climate Change Canada's estimates of emissions by provinces in 2030. Figure 5.6 shows these 2030 projected emissions for each province relative to the burden-sharing rules from figure 5.5. These 2030 projections take into account all measures in place in September 2019, along with all measures announced by that time that are known with sufficient detail – so the best possible case at the time as it included both actual and proposed policies (but not accounting for the proposed more recent federal policies such as the higher carbon price).

There are three groups of provinces. First, the projected emissions for four provinces – Alberta and Saskatchewan, and to a lesser extent Manitoba and Newfoundland and Labrador – are higher than they would be under most of the rules. Interestingly, the first three of these provinces have not set any emissions targets for 2030. Newfoundland and Labrador set a "proportionate" target – that is, the same as the Paris commitment.

Second, the projected emissions for five provinces – Ontario, Quebec, New Brunswick, PEI, and Nova Scotia – are essentially

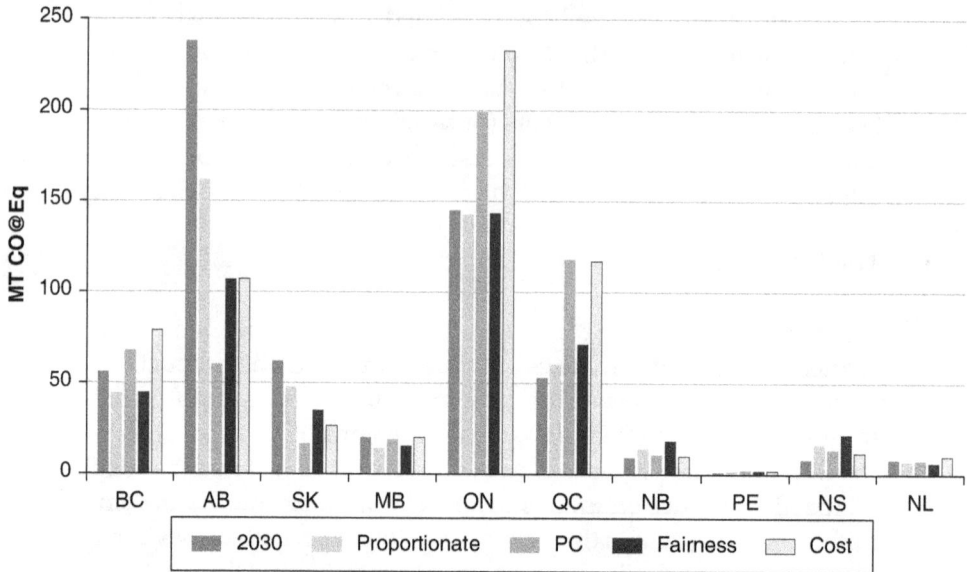

Figure 5.6. Projected 2030 emissions and 2030 emissions under various sharing rules. "2030" are projected emissions taking into account all measures in place and announced with sufficient detail in September 2019. "Proportionate," "Fairness," and "Cost" are as defined in figure 5.5.

Source: Canada, *Canada's Fourth Annual Biennial Report on Climate Change* (Gatineau: ECCC, 2019), tables 5.8 and 5.10; Statistics Canada provincial population data for 2017 and GDP data for 2018.

equal to or lower than what these provinces would be allocated under any of the burden-sharing rules. All these provinces have set targets for 2030. None aimed at a per capita approach or even explicitly for a redistributive approach. Quebec, PEI, and Nova Scotia targeted emissions greater than would be required under the proportionate rule. Ontario used to be in that group, but Conservative premier Doug Ford revised its target to the proportionate rule, stating that Ontario "should not be expected to do more than its 'fair share' to meet Canada's national commitment to reduce greenhouse gas emissions."[36] New Brunswick also has a proportionate target.

Finally, BC's projected emissions are higher than they would be under the proportionate or fairness approaches (it is a wealthy province) but lower than using the per capita or cost approaches. Its target is actually for greater emissions than would be allocated under the proportionate rule.

What does this all tell us? Almost half the provinces were not on track to meet any of the burden-sharing rules, and a move to any of these rules would hurt the resource provinces such as Alberta and Saskatchewan relative to the others. On the other hand, many of the other provinces would bear less of the burden if we moved to one of these explicit rules.

Before we talk about what we should do about divvying up the burden, it is useful to see what these differences in projected emissions mean for the Paris target. Some provinces such as Alberta and Saskatchewan would argue that they need higher emission allotments because they contribute more to the Canadian economy, which is good not only directly in the sense of providing jobs and revenue but also indirectly in the sense of government revenue for public services and reducing of emissions elsewhere. We will look at the government revenue argument more closely in chapter 9, but this approach has significant effects on meeting the Paris target. Alberta and Saskatchewan currently contribute about half of Canada's emissions, but if they emit GHGs as projected for 2030, they would account for almost 60% of the emissions allowable even under the former Paris target, rising to almost three-quarters under the new target. The oil sands alone would make up almost 20% of the former Paris Target by 2030. If the oil sands industry emits GHGs at this level, the remaining provinces would have to reduce their projected emissions by about one-third to make up the difference under the old target and even more under the new one. The federal government's promise to impose a cap on oil sand emissions would help, though note that baseline would at best be 2021 or 2022 oil sands emissions, with the rest of the country needing to make up for this exemption to meet the reduction from 2005 emissions.

In sum, the federal and provincial governments take a mixture of approaches, none of which seems sufficient. Many of the provinces now appear to be adopting the proportionate rule as their goal, but five provinces are projected to exceed the levels they would get under this rule. Moreover, for Canada to meet its Paris

Agreement targets and for Alberta to expand as it is proposing, the rest of the country will have to make drastic reductions. The situation is even worse if we think about trying to reach emission levels in 2030 that would limit temperature increases to 2 degrees C (as opposed to just Canada's Paris commitments) – Alberta and Saskatchewan already make up about 90% of that level of emissions.[37] And it is worse still if we try to hold to a 1.5 degree C increase.

How Should We Split It Up?

The current approach of minimal federal standards coupled with provinces setting targets on their own is rooted in the diffusion of responsibility across the federal and provincial governments as well as provincial ownership of natural resources. The federal government has taken a minimalist approach to national climate policy. The Pan-Canadian Framework was a step forward in that it attempted to get all the provinces pulling in the same direction, but they are not yet in sync. As a result, we have serially set and failed to reach our targets. Adopting policies we discussed in chapters 3 and 4 to address competitive pressures would help allay provincial concerns, but we still need to have some idea about how to share the costs. While there are a range of solutions that could be offered, we can divide them into four broad groups:

Keeping It Flexible. The status quo – no burden-sharing rule, each province setting its own goal, federal involvement relating to minimum pricing, limited public information[38] – has some benefits coming from flexibility and possibly even from a lack of transparency. This approach means both the federal and provincial governments can adapt in a rapidly changing world. Take competitiveness pressures. While much of the world seems on board with attacking climate change, the actual commitment of other countries (and in particular our major trading partner the United States) varies. The pandemic and the recent global situation have thrown up a whole other hurdle. Climate policies change as governments change and economies boom or bust. The flexible approach gives governments the ability to be nimble in the face of shifting competitive and economic conditions.

Moreover, allowing provinces to take the lead can result in efficient adjustment to local conditions. Provincial governments may be better able to understand the needs of their economies and design targets and policies to address them. It may also allow greater progress in the long run as individual provinces experiment with different policies. As we will see in chapter 10, these are common arguments in favour of decentralization.[39]

In addition, trying to be specific may mean trying to do too much and thereby doing nothing. Our governments may spend so much time and effort trying to make everyone move together that they do not move forward. The year 2030 would come and go and emissions would continue to rise. One benefit of the current approach is that it is a form of partially complete agreement – it may be enough of an agreement on reductions to move forward where a complete agreement would not be possible.[40] Our governments can agree that allocations should be "fair," but trying to get an explicit agreement on much more than that is too hard, if not impossible.

The best that can be done may be some common agreement on a direction and then working to find other ways forward. Common ground may be found, for example, on areas that overlap with climate policies, whether it be green economic development or health benefits from reducing emissions from coal.[41] In this way, the federal and provincial governments may be able to piece together a strategy that achieves reductions that would not be available if everyone had to explicitly agree to a strategy.

Making a Rule. If a discretionary, flexible approach does not seem to be working, what about instituting a formal rule about the allocation of costs? The obvious question is: who sets the rule? Chapter 10 discusses the Supreme Court of Canada's decision finding the federal government has the power to set minimum carbon pricing requirements. We will also look at whether the federal government has the power under the Constitution to order provinces to reduce emissions by a certain amount. If not, the provinces would have to be willing to come to some form of agreement. Political scientist Douglas Macdonald is one of the few who have studied the politics of this issue of burden allocation; he argues that even if the federal government does have broad powers, it would not use them without provincial agreement.

Macdonald has examined how the federal and provincial governments tried to find agreement on splitting up the burden in the 1990s but abandoned these discussions prematurely and instead tried to find easy sources of agreement.[42] As a result, he sees a lost opportunity to get movement on the costs from climate action, which he views as requiring a multilateral agreement on burden sharing.[43] The Parliamentary Budget Officer similarly has argued that the differences in costs across provinces undermine the ability to obtain some consensus on action.[44] Even the modest federal moves to set some common stepping stones have met with growing provincial opposition. The argument is that we need to agree on how to address these costs, as they will be allocated no matter what we do and will continue to be a barrier to more significant action.[45]

The most natural option may be the per capita rule, though this is also the option least likely to be agreed upon. While large provinces like Ontario would have easier restrictions, Alberta and Saskatchewan would face extremely high costs in meeting the targets. As has been argued about the per capita rule in allocating burdens across countries, "insistence on the first-best outcome, as a matter of principle, may make the climate change problem intractable," thereby hurting the most vulnerable.[46]

The proportionate approach seems more promising. Many provinces have already settled on this approach. For those with high emissions like Alberta and Saskatchewan, it is the least harmful of our allocation rules and seems most likely to secure their support.[47] Explicitly adjusting this rule for either efficiency or fairness only makes matters worse for the resource provinces and less likely to attract agreement.

An agreement is not outside the realm of possibility. The European Union, for example, used a "triptych" approach across states, looking at differences in economic structure, energy sources, and ability to pay (as measured by relative GDP per capita of member states).[48] A mixed approach could take into account both the differences in the costs of action and the fairness of different forms of sharing. Macdonald argues that the precision and certainty from a rule would be worth the delay and loss of flexibility, as all governments would be committed to the resulting allocation.

Paying for Change. What about having resource provinces pay the rest of the country to reduce emissions? If the claim is that the development of the oil sands is good for all of Canada, the profits from its development could be put back into other emission-reduction measures. This looks to be an aspect of the Trudeau plan – allow pipelines and oil sands development in order to obtain the money needed to address climate change. The European Union partially relied on financial compensation to help allocate the costs of reaching its target across its member countries, by providing a percentage of the funds from its emissions-trading system to countries with fewer resources to allocate to emissions reduction.[49]

This argument is related to one version of the "polluter pays" approach. Alberta, for example, has had a tremendous run, benefiting greatly from oil booms in the past. It had huge employment gains and attracted young workers to the province, and in fact has done much better than other provinces over the same time period.[50] It was able to raise public spending without increasing taxes. This version of the polluter pays argument would be that Alberta should use some of those gains to pay for the clean-up.

In principle, such a plan would provide significant funds for emissions reductions and, in that sense, it seems rational to use any revenue to fund other reductions.[51] However, there are two big concerns. First, under the Constitution, it is not clear how provinces such as Alberta or Saskatchewan could be compelled to make such payments beyond increasing the carbon price. Carbon price revenues, as they are based on emissions, are unevenly distributed across the country and could be used to make payments to provinces for emissions reductions.[52] The oil price crash and economic downturn related to the pandemic makes a large-scale, explicit attempt to redistribute through the carbon price seem a political non-starter.

Second, even if it was possible, and granted that increased production of fossil fuels would provide revenues for reductions in GHG emissions in other provinces, it would not only increase Canadian production emissions but also contribute to emissions from the use of the oil and gas. Some argue that to stave off dangerous climate change, oil needs to stay in the ground. Others, conversely, point to the potential "Green Paradox" – that reduced production from the Canadian oil industry would merely be replaced by other, less clean/less moral sources of oil, and we should focus on developing cleaner production technology that can reduce emissions around the world.[53] The trade-off would be (and is) controversial even without the added hurdle of oil price uncertainty. In part the use of carbon pricing and subsidizing of carbon capture technology is intended to reduce the costs of continued oil sands production through reducing at least the direct emissions.

Paying Off the Losers. National agreement or sharing could, conversely, be built more directly. If the country as a whole needs to hit a particular GHG target, and that target is going to be particularly burdensome for some provinces, then those provinces should be compensated. If the emissions from Alberta and Saskatchewan are too high and the Constitution gives them the power to control their own resources, they could be paid for their losses.

The federal government is doing this now to some extent. Take the shift away from coal: the federal government set national standards given the importance of this source of emissions. It also, however, has given subsidies for the shift and is compensating communities that rely on coal-based activities. The plan is for provinces that have not done so to get out of coal, and the federal government is attempting to share the costs of this move nationally.

We can also see this buying-off strategy to a lesser extent with the federal government's carbon price program: either the provinces impose a carbon price or the federal government imposes the price and returns the revenues to the province. In either case, the costs of transitioning stay within the particular province.

Paying off those who are hurt is in a way a straightforward way out of the problem. The costs are spread more widely, which hopefully brings buy-in from those who are most impacted.

The difficulty, of course, is that those costs still have to be paid. The Trottier Institute in Montreal has argued that federal and provincial policies in 2021 are not sufficient to meet Canadian reduction commitments and that reducing oil and gas production is necessary.[54] If Alberta or Saskatchewan leaves the oil in the ground, they bear a huge opportunity cost in terms of lost revenue. If the rest of Canada attempts to offset these costs, the actual outlay of resources may be too large to be politically feasible.[55] Some other jurisdictions, such as the European Union, Australia, and Germany have attempted to use financial compensation,[56] but the scale of such a measure in Canada in terms of oil is huge.

The difference in costs of climate change across provinces is perhaps the greatest obstacle to reaching our Paris goal, let alone undertaking more ambitious reductions. We need to find common ground on who should be doing what. Should increased production of oil and gas be allowed with greater payments made towards other emissions, or should oil and gas production increases be stopped but with the resulting costs borne more fairly? Our governments have done little to come to a mutual understanding.

Douglas Macdonald claims that the failure to explicitly deal with this barrier has been a significant cause of Canada's lack of movement.[57] It was not as much of a political issue in the past, as governments did not really take their targets, or the national target, seriously. As they have begun to do so, there has been greater conflict. For Macdonald, the way forward is to first explicitly take on the allocation of the costs of climate change: "If we start with easy issues, governments are likely to agree on those and then stop there, avoiding the hard issue of allocation in favour of the much more pleasant task."[58]

Macdonald maintains that the outcome will only be accepted as fair if it comes from a fair process. Part of the answer for him is to get beyond unilateral policymaking and work towards true intergovernmental agreement. He goes further than burden sharing and argues for the federal and provincial governments negotiating new national targets with the provinces setting new targets based on this national target. Such a process, he claims, would bring legitimacy to the targets and result in a goal to which all will feel committed.[59] In its most recent climate plan, the federal government has espoused cooperation and indicated it will seek provincial input on new national targets. As we will see, however, the Supreme Court has strengthened the federal government's hand, and the latter continues to make plans at times seemingly unilaterally.

We will come back to dealing with this issue of cost sharing more concretely, but first we need to look at a third institutional concern – too much deference.

chapter six

Deference: Where Are the Guardians?

So far we have seen how Canadian environmental law relies on broad legislation that grants discretionary powers to Cabinet or others to make policies and regulations, to approve projects, and to make plans. This pattern holds at both the federal and provincial levels. Moreover, power over environmental issues is diffused across federal, provincial, and territorial governments. None have exclusive powers to make environmental laws, and each plays some role in addressing problems like climate change or species at risk.

Reliance on discretion and diffusion has its benefits. It allows flexibility and provides an opportunity to rely on expertise and local knowledge and to experiment. At the same time, however, it affords space for narrow interests to influence what should be policies aimed at the public interest, and allows for finger pointing and the avoidance of action. We have seen the results in the weakness of Canadian climate targets and ineffectiveness of climate policies.

How do you get the benefits of discretion and diffusion and minimize their downsides? The answer lies in both legal and political institutions. Before we can talk about the solutions, we need to get a better grasp on what has been going wrong. The central issue has been deference – too much trust in political accountability by voters and by our courts. In a way, Canada has had the opposite problem to the United States, where regulatory laws and policies have been driven by a lack of trust. US Supreme Court justice Stephen Breyer has written about the "vicious circle" of US policy on dealing with risks such as to the environment.[1]

Public perceptions of problems influence congressional reaction in the forms of laws, which in turn distorts regulators' policies – the result is inefficient, irrational, and unfair law and policy. At the centre is a lack of trust in regulatory decisions. The Canadian problem has in the past seemed to be the opposite – too much public trust that government will make rational environmental laws, with the result that there is less public scrutiny and those who care a lot (generally industry) have had considerable influence in how laws are shaped.

Expertise, Accountability, and Law

To understand the concern about deference and trust, we need to think about what goes into the belief that governments are making legitimate decisions. Three broad arguments justify government regulating in any area, including climate change: (1) the decision is based on the best available evidence and knowledge – that is, it is based on *expertise*; (2) the decision has been decided on in a fair and inclusive fashion – its roots are in *political accountability*; and (3) the decision is in line with what we feel the *law* demands.[2] A quick note on terminology – "government" is often used to encompass both the elected officials making the laws (such as a law that permits carbon pricing) and the Cabinet, ministers, civil servants, boards, or agencies carrying out what the law requires (the actual details of how the carbon price will be implemented). As we noted earlier, our main focus is the latter (in legal terms, the "executive"), the ones who implement the laws. Both making and implementing laws involve questions about whose values matter and the relative power of different parties, but the use of discretion granted by legislation is particularly pernicious in the area of implementation.

Expertise

One of the first things that comes to mind about what makes a good decision is that it be based on the best possible information and knowledge. We want decisions on the appropriate interest rate to involve people with financial and economic knowledge and with access to good data. If we are deciding whether to allow a chemical to be used, scientists should be examining the risks to health. Decisions on how to reduce greenhouse emissions will require the knowledge and expertise of scientists, engineers, economists, financial analysts, and others. How can we make a reasonable decision on complex matters without relevant information and applying what we have learned about the world?

We will refer to this as "expertise," but it should not be defined narrowly. It includes involvement of people generally thought of as "experts" in particular areas like economics or physics or neuroscience, but to be most effective, the term must also include other forms of knowledge. Indigenous knowledge is a clear example. Indigenous knowledge and experience encompass understandings of the environment that are different from and sometimes conflict with that of more mainstream sciences. Canadian governments are struggling to find ways to incorporate both forms of knowledge. Expertise can also be gained from experience as opposed to study – experience in a job or even more basic life experiences in some cases, such as experience of discrimination.

For a decision to be based on expertise, the expert has to be independent to some extent from influences that inappropriately diminish the weight of that expertise. Much of Canadian law delegates power to decision-makers because of their expertise; examples are the Canada Energy Regulator or the Canadian Nuclear Safety Agency. Independence on this view would mean that their decisions are not inappropriately swayed by politics or interest groups. They should be able to draw on knowledge of energy markets or nuclear safety. We hear this sentiment in the cries that we know what the science says about climate change and what to do to reduce emissions, and we should just do it. It taps into a desire to find rational solutions. As we will see, expertise cannot easily be separated from either political accountability or the law, but as a concept it has powerful sway.

Political Accountability

The difficulty with making decision-makers more independent so they can use their expertise is that they become less politically accountable. Political accountability can make us feel decisions consider our interests as people – in economist Amartya Sen's terms, as agents as opposed to "passive recipients" of "cunning" policies.[3] At its best, it allows us to make value trade-offs in a manner that considers what Canadians want and care about. Accountability can come from the fact that ultimate power is wielded by elected officials. Our elected legislators make laws and they delegate important decisions to Cabinet (either federal or provincial), which is also composed of elected representatives. But political accountability and democratic legitimacy also can come from other bodies, depending on how they are set up. For example, a government panel assessing the environmental effects of a mine may lack legitimacy if it is composed solely of mining industry representatives, but can gain legitimacy by having members from other backgrounds or through broad public participation processes.

Accountability is not good, however, when government officials are accountable to special interest groups that sway policy towards their own interests as opposed to the public interest. Political incentives can lead governments to make decisions that benefit more powerful groups at the expense of minority interests. Canada's treatment of Indigenous peoples is an obvious and ongoing example, but there are many more. Accountability can also be bad when the public is mistaken (such as about the probability or nature of harm from climate change) but public officials listen to them anyway.[4] The public has for some time not seemed to recognize the risk of climate change, and public officials responded to (although also fed) that lack of concern.

Accountability has a difficult relationship with expertise. Many decisions are not straightforward applications of expertise, and in fact you could argue most are not. Nuclear safety might seem to fall into the category of a decision that should be made by an expert. We want to be sure that nuclear facilities are not going to melt down and endanger lives. Yet even here science cannot give the whole answer. A few years ago the Nuclear Safety regulator wanted to shut down reactors at Chalk River Ontario for safety reasons, but the government wanted them kept open as they were an important source of materials for specific medical treatments.[5] Unlike the experts, the government focused on trading off the safety risks with the benefits to health. In the climate context, we need a way of deciding how much we are willing to trade for lower risks.

So at its best, political accountability allows us to make value trade-offs in a manner that considers what Canadians want and care about. At the same time, however, we need to worry about accountability's downsides – that it may lead to decisions that harm those with less political power, that its reliance on public input may increase harms if the public is mistaken, that it may allow interest groups to divert decisions away from the public interest.

Courts/Law

A third way in which we can gain confidence in government decisions is if they are made in accordance with the law. The concern here is not so much about individual laws, but about how we make those laws and policies. This ties into the idea that the courts need to make sure the decisions are legitimate.

For climate change decisions, we can think of three broad issues. First, decisions must accord with the basic rights Canadians have under the Constitution and the Charter of Rights and Freedoms. Decisions cannot undermine Canadians' right to life or security of the person or to be free from discrimination. The courts are the guardians of these rights. Relatedly, the Constitution includes rights of Indigenous peoples. Because of their particular importance, we will talk about those separately in the next section.

Second, laws and policies have to fit within the basic federal structure of Canada. We talked about this issue in the last chapter. The federal government cannot impinge on provincial powers and vice versa. At times this constraint seems like an obstacle to good policy, particularly in addressing climate change, but we will discuss how it can actually be beneficial. Again, the courts are the prime protectors of this constitutional structure.

The last element is the most vague: our decisions should be in line with what lawyers call the "rule of law." At its most basic, the rule of law requires that government decisions be made following a fair process and in most cases be at least reasonable if not correct. Government decisions are made by a wide variety of actors – Cabinet, ministers, department officials, boards. As we have seen, elected officials make laws that delegate powers to these actors to make decisions because these actors have greater expertise, have more time and information to make a wide variety of decisions, and can decide more quickly and possibly fairly. How do you ensure these actors are doing a good job and not either making mistakes or using their powers to further their own interests?[6]

Unfortunately, the Supreme Court has struggled to bring "greater coherence and predictability" to the role the courts actually play in this process.[7] It tries to ensure that judges respect the fact that the legislature has delegated to some body (Cabinet, the National Energy Board, a government official) the power to make certain decisions, but at the same time that judges ensure the fairness and rationality of those decisions. Central for the Supreme Court is that government decisions are supposed to result from a "fair" process and be "transparent, intelligible and justified"[8] – easy to say, really hard for judges to figure out what that means when reviewing the extremely wide variety of government decisions. It means something different for an immigration officer deciding on someone's application for admission to the country than it does for a National Energy Board panel deciding whether to recommend the Trans Mountain pipeline. In general, judges are not supposed to consider how they would resolve the issue (whether the person should be admitted or the pipeline approved), but whether the process for making the decision was fair and whether anyone challenging the decision has shown that the decision is unreasonable.

For judges to figure out whether the decision was reasonable, they need to look at the reasons given and determine whether the decision is logical and justified in light of the facts and the law. The key underlying concern is how deferential the court is to the government. Think about what a judge should do if an individual challenges a government approval of a pipeline on the basis that the government failed to adequately account for the adverse effects of a spill. The more deferential the judge is, the less she is a check on the fairness or reasonableness of the decision; the less deferential, the more the judge gets to decide whether building a pipeline makes sense.

Judges have the benefit of being largely independent of direct political influence,[9] but obviously have their own limitations. They are generalists – they have no special expertise in areas such as environmental protection. One of the key reasons why we may want judges to defer to government decisions is a recognition that the legislature gave power to someone – the Canadian Energy Regulator, the CRTC, immigration officers – because they have *expertise* in a particular area. This expertise puts them in a prime position to interpret what phrases like "adverse effects" mean in environmental legislation or "predatory pricing" is in the context of the *Competition Act*. The courts at least in theory take this limitation into account through a deferential approach.

The other principal limitation of courts for our purposes is that they are not politically accountable. If an act specifies a pipeline may be approved if it is "in the public interest," judges are not in any special position to decide what that means for Canadian society and how different interests and values should be traded off. Acts tend to give a minister or Cabinet or some other decision-maker such broad discretion because they are at least somewhat politically accountable. As a result, the courts take (or are supposed to take) an even lighter hand on, and be reluctant to substitute their views of, such vague, *value-balancing* decisions.[10]

Indigenous Rights. Before we leave the issue of law, there is another part of the Constitution that is important to our discussion of climate change – the section 35 rights of Indigenous peoples. Section 35 recognizes and affirms "existing aboriginal and treaty rights of the aboriginal peoples of Canada." The Supreme Court of Canada has developed an extensive framework for deciding when decisions by the federal or provincial governments or others do not respect these rights. Indigenous rights have been raised in a wide variety of contexts, from fishing and hunting to the impact on Indigenous peoples of large projects such as dams, mines, and pipelines.

For our purposes, a key element of this constitutional right is that the government has a duty to consult and accommodate Indigenous peoples before doing something that could adversely affect a potential Indigenous claim or right.[11] The Court has taken a very broad view of when the government has this duty – essentially any time it contemplates something that might impact rights of which it is aware. The government must ensure that the process is consistent with the "honour of the Crown."

The duty has led to considerable litigation – most famously with Indigenous groups going to court on the basis that the federal government did not adequately consult them or accommodate their concerns in approving pipelines from the Alberta oil sands to the coast. Part of the difficulty lies in what is expected in these efforts to consult and accommodate. Consultation includes some basic elements like notice that the government is going to act, providing information to the affected peoples, and giving them the opportunity to respond (including potentially providing funding). It also at times includes more action on the part of the government, such as ensuring that someone is at the table who can actually make decisions.[12]

The courts do not demand perfection, but only reasonable efforts by the government to consult and accommodate. Given the number of parties and complex issues generally involved, the Federal Court of Appeal has noted that "in attempting to fulfill the duty, there will be difficult judgment calls on which reasonable minds will differ."[13] Further, the Supreme Court has been very clear that this duty does not give Indigenous peoples a veto. The scope of the duty may change with the federal government's legislation implementing the UN Declaration on the Rights of Indigenous Peoples with its reference to the "free, prior and full consent" of Indigenous peoples, though what that means in terms of a veto is still to be worked out.[14]

The Constitution then requires inclusion of Indigenous peoples in decision-making. Given the nature of what needs to be done to address climate change, this duty will be important in a range of situations, most obviously on projects that continue to extract and move fossil fuels, but also dams and windfarms or transit projects. The Constitution requires cooperation across a range of governments, with the inclusion of Indigenous peoples increasing in importance exponentially.

The Three-Legged Stool

We therefore have these three bases for legitimate government decisions: expertise, political accountability, and the law. They are interconnected and all three are necessary. Expertise, for example, often is spoken of as objective and apolitical. However, expert knowledge is filled with and shaped by value judgments – such as about what is an acceptable level of risk, what research is necessary, who gets research funding, and which research gets promoted. This connection underlies a range of issues, including not only environmental impacts but assessment of the costs and benefits of climate action. Why do we care about the environment? We will discuss how participation of the public generally and Indigenous peoples can provide evidence of both the harm from climate change (such as the cultural and welfare implications of environmental harm) and the values of those who will be affected. Further, the version of the "rule of law" set up by the courts (that we want courts to ensure regulators have legal authority for their decisions, follow a fair process, and act reasonably in light of the legislation) ties into a narrow version of what we feel are legitimate constraints on regulators, leaving them with a fair amount of space to exercise discretion.

The weaknesses of each of these bases for decisions are balanced by the strengths of the others. Expertise provides knowledge and experience for better decisions but lacks accountability if the decision requires some form of value judgment, or may lack fairness or independence depending on what the law says about how decisions should be made. Political accountability can give legitimacy to value judgments but may not bring knowledge or expertise, such as about the potential effects of the decision, and may be subject to problems such as capture. The law and courts can provide a solid framework in fairness but judges may lack the expertise necessary for good decisions or the accountability necessary for value judgments.[15]

It is like trying to sit on a three-legged stool. The stool will not work if a leg is missing or is much longer (or shorter) than the other two. There is no perfect length of leg, but the stool requires a balance in the support of all three legs to function. Adrian Vermuele, an American legal scholar, argues that each of these bases has its supporters and that we tend to continually swing between the three, with the result that there is "no stable core, no equilibrium solution."[16] However, while we will never find a perfect balance, there is a core, or basic balance, we can stay within and know when we are off it. We become concerned about making sure policy is based on science but grow disappointed when science does not allow us to make trade-offs across different groups in a way they accept. We then turn to greater political accountability only to become disappointed when the decisions do not rest on science and are subject to the power of well-resourced interest groups. We try to solve this by giving greater powers to the courts to straighten things out – and so on.[17]

When things have gone too far, we need to adjust each leg to come back into the basic balance that serves the core function – to provide efficient and fair decisions that are seen to be legitimate. We will only get acceptance when people trust that the system is making good, fair decisions, and such trust requires us to keep in balance.

Tilting Badly

The obvious question is: are we in balance now? Certainly, if we think about the level of confidence in government, something seems wrong. While each leg has problems, the main concern is over-emphasis on political accountability – and in particular the willingness to leave so much to discretionary decisions of various government bodies. As we discussed in chapter 3, discretion has always been a problem for environmental law, and it is at least as much if not more so for climate change law. Climate laws rely on grants of discretion for setting how emissions should be priced, who gets transitional relief, when methane regulations kick into force, whether a new pipeline should be built, and so on.

We need to ensure that decisions are made in the public interest. Industry may sway decisions away from the public interest through threats about shutting down or moving due to the costs of climate change, or through its resources or information to coax favourable decisions. Again, climate change particularly raises these concerns given Canada's dependence on both the production and use of fossil fuels, leading to worries that not only the economy but the political system and laws were structured to foster these industries.[18] Legal scholar Jason MacLean argues that "Canada's climate policies are largely about climate in name only. They are designed, instead, to further the special interests in continued oil and gas extraction and export."[19] Alternatively, decisions might respond to irrational public reactions.

To see the difficulty, we need to look at the role each leg is playing in controlling the exercise of discretion:

Political Accountability. In terms of political accountability, an obvious check on discretion is the public. We rely on elections to provide the most important sanction to elected officials, including the prime minister, premiers, and Cabinet ministers. They are responsible for their choices concerning the environment as well as those of civil servants or bodies they create, like the Canadian Energy Regulator or the Impact Assessment Agency.

Accountability may also be enhanced by increasing the role of the public in decisions. Environmental decisions such as whether to limit emissions of methane or to approve a new mine used to be mainly the product of discussion between government officials and industry.[20] The government was assumed to fully and faithfully represent the public interest, so there was no need for involvement of the public. Beginning in the 1970s and 1980s, however, both the federal and provincial governments began to open up their decision-making processes, such as through providing public notice of and the ability to comment on proposed regulations or approvals.

A prime example is the approval of potentially environmentally harmful projects like mines, dams, or pipelines. Following the lead of the United States, in the 1970s the federal government began developing processes for assessing such projects. These environmental assessment processes attempted to bring the public into an examination of the environmental impacts of these projects before they were built. They drew on expertise through having panels of experts undertake extensive public reviews of projects and engage the public generally and Indigenous groups in particular in the process. As political scientist Mark Winfield has stated, "In Canada, the central political rationale for establishing environmental assessment processes was to provide structures through which social and political conflicts over the distribution of costs and benefits associated with resource and infrastructure projects could be addressed in a manner that participants would regard as procedurally just and fair, and therefore legitimate in their outcomes."[21]

Transparency helps. Panels review applications and hold public hearings. They issue a public report that sets out their findings and their recommendations. Cabinet then has to decide whether to give its approval or not. The hope is that such a combination of expertise and public input going into the panel process and political accountability for the Cabinet decision will lead to a reasonable decision that is seen as legitimate.

Political accountability of this sort depends on the public, but the public is notoriously poor at holding public officials to account, and industry is particularly good at it. A 2018 poll found that there was a fair level of concern about climate change in Canada, and that concern exists across the political spectrum – with almost 60% of "right"-leaning people and 80% of "left"-leaning people seeing climate change as a major threat.[22] But public attention can be sporadic and easily lost. As we discussed, attention to environmental issues has tended to go in waves, declining when economic issues become of central concern. The environment dropped from people's focus with the advent of the COVID-19 pandemic and the resulting economic downturn, though there remains considerable support for climate action.[23]

Such a check on discretion also requires that the public is informed and "rational." The public gets their information from the government but also from the media, and the media may be skewed in their presentation or interpretation of the facts and the problem.[24] As well, some, such as the fossil fuel industry, spend considerable sums of money to promote their views.[25] And as we will discuss in chapter 11, people make systematic mistakes when dealing with issues like climate change that involve complex, unfamiliar issues and small risks of catastrophic harm.

Transparency and political accountability through both the electoral process and more direct involvement in decisions is important but has so far been insufficient to promote climate action that is sufficient or viewed as legitimate. Hope is growing that greater public concern about climate change will lead to greater action. Some such as Jaccard feel the answer lies in electing "climate-sincere" politicians.[26] While the choice of elected officials is central, the concern is that the public will not sufficiently become engaged in the near future to hold politicians to account.

Expertise. The concerns about grants of discretion and reliance on political accountability might not be a problem if the other two legs of the stool could be relied upon. However, both are limited. Expertise could put bounds on how discretion can be exercised, even if it cannot be relied on for value judgments, and it does inform climate decisions. The difficulty is the extent to which it is subordinate in major decisions to the discretion of political officials.

Going back to our example of environmental assessments, the National Energy Board (now the Canadian Energy Regulator) reviewed the application to build the Trans Mountain pipeline. The *Canadian Environmental Assessment Act* (now the *Impact Assessment Act*) required the board to take into account science and traditional knowledge. But under the act, the board only recommends a solution to Cabinet, which makes the ultimate decision – and the board itself is appointed by Cabinet.[27] The transparency of the process helps us feel more comfortable with the political accountability, but it is clear that expertise is subject to political accountability. Winfield argues that the environmental assessments process eventually lost some of its ability to provide that legitimacy as the government came to be seen as promoting projects rather than assessing them.[28] Discretion was seen as aiding industry interests as opposed to furthering the national interest. At the same time, industry decried the uncertainty of the process. Such primacy of political accountability and discretion is found throughout climate law.

On a larger level, while Canadian governments at least try to mix expertise and discretion to decide one-off projects like pipelines and mines, they lack a good system to use expertise to decide bigger questions such as whether the pipeline or mine fits with Canadian climate change goals. Having Cabinet approve projects is supposed to allow these decisions to be made in the public interest – but there are no criteria on which to base these decisions or to hold politicians to account. As a result, these individual project decisions became "proxy venues for debates over the future role of fossil fuels in Canada's economy."[29] Environmental assessments became battle grounds for debates that were not being held elsewhere. In chapter 11 we will discuss more recent attempts by the federal government to increase the positive connections between political accountability (including targets and public reports) and expertise.

Courts. What about the third leg – the courts? As we saw, courts can support the legitimacy of environmental decisions by ensuring they comply with constitutional obligations and are consistent with the "rule of law" more generally. In terms of the Constitution, the Canadian Charter of Rights and Freedoms provides Canadians with such rights as to life, liberty, and security of the person as well as freedom from discrimination. Both have been used, or are being used, to try to get government to act on environmental issues such as air pollution affecting the Aamjiwnaang First Nation, an Indigenous community near Sarnia, or mercury poisoning of the people of the Grassy Narrows First Nation.[30]

Groups are also attempting to use the Charter to hold governments to account for failing to address climate change and meet climate change commitments. Such challenges face obstacles, including convincing a court to second-guess the government in its decisions balancing these rights against other public interests.

For example, the Federal Court found that a Charter challenge to the federal government's climate plan was not appropriate for the court to decide as it was too broadly framed and, in any event, had no reasonable prospect of success.[31] A challenge in Ontario, on the other hand, that was more narrowly framed has been allowed to proceed.[32] So far, courts have been reluctant to use the Charter to support challenges to weak environmental laws.

The courts can also give content to the Crown's constitutional duty to consult and accommodate Indigenous peoples. Beyond the issue of the lack of a veto, the courts' approach to the duty to consult has raised concerns about its value for Indigenous peoples in its current form. It has been argued to be overly deferential to government decisions and legislation, narrowing the applicability and scope of the duty.[33] Further, the Supreme Court has noted that its aim is to foster negotiations and reconciliation rather than continual litigation.[34] That has not worked well so far on some big issues, such as the TransMountain Pipeline, which has been the subject of continual litigation about whether the federal government engaged in sufficient consultation and accommodation, as were earlier pipeline proposals.

The other main area in which courts could play a role is through individuals challenging the fairness or rationality of a particular decision. What has been clear over the last decade until very recently is that the Supreme Court has increasingly asked judges to be more and more deferential. As the Court notes, this does not mean that when a decision is challenged a judge should blindly agree with it. However, on balance the approach has been to give government a wide space in which to roam. Legal scholar Jocelyn Stacey sees the courts as responding to discretion by creating "black holes" and "grey holes."[35] The black holes are decisions that the courts will not review or will review only on a very minimal basis – because, for example, the court claims the issue is too political. She points in particular to the basic unwillingness of courts to review the content of regulations or the process by which they are made. Grey holes exist where the court seems to be willing to review decisions but in effect does not. Stacey uses the example of an act that gives someone power to act "if satisfied" of something – such as if the decision is in the public interest. The legislature can set out criteria in advance, but courts are unwilling to review these decisions unless, for example, the decision-maker had no evidence on which to base the decision.

The problem is that discretion is needed to give flexibility and allow expertise to work, but for courts to act as a check on this discretion they must be willing to review the decisions.[36] In part, courts are concerned about being drawn into scientific debates or debates that involve trade-offs.[37] In its latest attempt to simplify and find a consistent approach to the courts' role in reviewing government decisions, the Supreme Court of Canada reaffirmed the need for deference but developed an approach it terms "robust" review based primarily on the reasons given by the decision-maker.[38] Two Supreme Court justices (Justices Abella and Karakatsanis) claim the Court's new approach "dramatically reverses course – away from this generation's deferential approach and back towards a prior generation's more intrusive one."[39] Legal scholars are divided on whether this is actually the case, and time will tell if this is true.

The danger from a strong (less deferential) approach to review is that the courts will decide the issue for themselves even though they claim to be deferring.[40] Justices Abella and Karakatsanis warn that "the rule of law is not the rule of courts."[41] The Supreme Court's most recent approach opens the door to judges either making mistakes from a lack of expertise or inappropriately imposing their own views of what a value trade-off should be.[42] We seem far from that point right now, however, with courts at times deferring so much that for some discretionary decisions they seem to play little if any policing role. The question will be whether the new approach, with its strong emphasis on reason-giving, can find the balance between these two polar approaches. We will return to this issue in chapter 10.

Getting a More Solid Footing

It seems we have a problem. We rely heavily on discretion, ostensibly in part to allow for expertise to play a role in decision-making, but we have set up the system to rely excessively on political actors with too small a role played by expertise, the courts, or even the public. The public has tended to trust, or defer to, government decisions so much that they have not played much of a constraining role on political actors. This deference is exacerbated by the diffusion of responsibility across governments, which makes it difficult for the public to determine who actually is to blame. The courts have also been quite deferential, limiting their own policing role over discretionary decisions. The result has been space for short-term decisions that have failed to allow us to take strong action on climate change. Discretion and regulatory capture lie at the core of the problem, as does their impact on the ability of expertise, the public, and the courts to counteract the power of industry.[43]

Is there a way out? The key is to find a way to build trust – not the kind that allows people to switch off and the public interest to lose out to private interests, but the other kind that engages people and gives them confidence that sound, reasonable decisions are being made. We need to work on all three legs of the stool for this to happen.

chapter seven

Focusing on People

The last few chapters looked at three main institutional elements that have hindered climate action: discretion, diffusion, and deference. Each, however, has its positive aspects that can be used to find a better path. We now turn to how to do so. We will start with the key to change – people.

There is a stark divide in climate politics. On one side are an array of politicians who are fighting strong action on climate change. They do so not, they say, because they deny climate change is happening or because they are against reducing emissions, but because the proposed measures will hurt working families. On the other side are the "new socialists" who see the solution to climate change and inequality as intimately linked and the whole economic and political system up for renewal. The most visible form of this argument is the Green New Deal, the grand plan in the United States to address everything from reducing GHG emissions to guaranteeing a living wage, providing health care and housing, and improving access to education.[1] The Green New Deal came to Canada shortly after, though the Canadian left had a version of such a plan earlier on called the Leap Manifesto.[2]

Motivating this debate is the impact of climate action on people. It will affect us all as individuals but to different extents, such as through increases to costs of driving or of heating your home. Remote northern communities may be particularly vulnerable to monetary costs from climate action because they face a disproportionately high cost of transportation for their goods, bear high winter heating costs, and in some areas rely on diesel generators for electricity. And these costs may take up a larger share of expenditures of families that have less income. Moreover, climate action may harm workers in industries that have to reduce emissions. Of course, the costs of non-action are also unevenly spread.

Part of the reason Canada has not seen consistent political pressure for change is that our governments are not dealing well with these different impacts of climate policies. It is another institutional story – our policies have focused importantly but narrowly on markets and economic growth. Both to ensure the fairness of the change and to build support for the transition, a broader lens is needed.

There is a theory of how regulations get passed that political scientist Bruce Yandle called the "Bootlegger and Baptist" coalition. The idea is that in the days of Prohibition, laws banning sales of alcohol were enacted because they were supported by two groups – those who wanted the laws out of principle (the Baptists) and those who supported them out of self-interest (the Bootleggers).[3] In the environmental area, a similar dynamic can be seen in the successful fight to regulate ozone-depleting substances. Strong action occurred when those who were concerned about the environmental harm of these substances aligned with those who benefited from their restriction (such as manufacturers of refrigerants that were not harmful to the ozone).[4] To get support for climate action, or at least reduce opposition, both groups are needed – those who want to avert the damage from climate change and those who stand to gain from the transition or at least do not fear it. Canada's current approach seems to downplay the concerns of those who, with some reason, fear that the costs of change will be visited on them. It limits the support for stronger action.

This chapter starts the discussion of how fairness and pragmatism point to the need to broaden our focus to build support for change. To understand how to reduce barriers to change, we first need a better idea of who actually may bear the costs from climate policy.

Who Will Pay?

We will look at two broad, overlapping divides in Canada: those
with more money versus those with less; and those employed in
sectors hit hardest by climate policies versus everyone else. There
are others that we could think about, such as whether carbon pricing may have disproportionate impacts by gender. We will not
explicitly discuss such differences, though they are obviously
important.

Rich versus poor. "Rich versus poor" is a poor way to capture this
issue, but it is the one that has been latched onto by some groups
who are arguing for significant change. We need to look at how climate policies impact people across the range of income groups in
Canada. First, it is important to note that even before we consider
the costs of climate policy, Canada has an inequality problem. The
inequality in people's "market income" – what they are paid for
their work before they pay taxes or receive government transfers –
has jumped over time, particularly in the early 1980s and 1990s,
and has remained high in recent years.[5] However, once we take
into account taxes and transfers, inequality is reduced considerably – by more than 28 percent. In the 1990s and early 2000s, this
after-tax and transfer inequality rose considerably before stabiliz-
ing (at the higher level) in the late 2000s.[6] In this pattern of rising
income inequality in the 1990s, Canada is similar to other developed countries, and is essentially average.

Much of the growth in income inequality stems from the fact that
while incomes for most people have not grown significantly since
the 1980s, incomes for the top 10%, and particularly the top 1%,
of income earners have gone up dramatically. At the lowest end
of the income scale, levels of poverty have declined or remained
stable overall in Canada since the 1980s.[7] There are also differences
across provinces, with Quebec and the maritime provinces having
the lowest median after-tax incomes and Alberta, Ontario, BC, and
Saskatchewan the highest.[8]

There are many possible explanations for why income inequality has gone up over time – technology may have led to some
jobs being lost or "globalization" may have caused jobs to go

overseas – but one underlying issue is of particular concern to us. The growth of the oil industry in Alberta may have raised the income of low-skilled workers in the oil and gas sectors as well as other industries – that is, the oil and gas industry may have kept income inequality from rising even further.[9] It is interesting, however, that at the same time those in the top income levels were disproportionately benefiting from the oil and gas sector.

The pandemic did not help in terms of inequality. Once the pandemic hit, unemployment grew by about three million and real GDP contracted by about 8%.[10] The harm has not been evenly spread; the economic effects of the pandemic appear to be worse for lower-income households and for women and visible minorities.[11]

Our concern is whether the costs of carbon policy disproportionately fall on the less well off, worsening this inequality. Individual Canadians face the costs of carbon policy through two broad avenues (leaving aside for now investments and any taxes needed to fund government expenditures such as on adaptation).[12] First, Canadians use fuels every day in heating their homes and powering their cars. Carbon pricing or regulations that impose a cost on fuels directly raise the cost of doing these things. Second, Canadians buy or use things that were made using fuels. The makers of these things face higher costs because of climate policy and may pass those costs on to consumers.[13]

We can think of different factors that may tie into what individuals pay for climate policies. If you live in a region that has more carbon-intensive production, the products you buy will likely be more expensive.[14] The emissions intensity of electricity production is an important factor in the cost of climate action for a province.[15] Ontario, for example, benefits from a low-carbon electricity system while Saskatchewan overwhelmingly uses carbon sources to generate electricity.[16] The difference shows up in higher average costs households will have to pay from, for example, a carbon price. Another obvious factor is household income. Not surprisingly, households with more income are likely to pay more as a result of the federal government's carbon tax, as they spend more.[17] Top income earners may pay more than twice as much as the lowest (before taking into account any rebates).

What is the overall impact of this? Economists refer to a "regressive" tax as one where taxpayers with lower incomes bear a greater tax burden. A "progressive" tax is the opposite: taxpayers with higher incomes pay a larger proportion of their income in tax. Climate policy may be progressive or at best only mildly regressive, even before we take into account measures to reduce the impact on low-income people. The BC carbon tax, for example, is mildly progressive even without any of the revenue being returned to low-income households.[18] More generally, the Ecofiscal Commission found that a $30 per tonne carbon price would have only modest average household costs (for most, less than 1% of total income) in Alberta, Manitoba, Nova Scotia, and Ontario.[19] The pricing is slightly regressive in all cases, though slightly more so in Alberta. Costs are higher in Alberta and Nova Scotia in large part because they both rely to a greater extent on fossil fuels to generate electricity.

Finally, the composition of different income groups varies.[20] Some groups – Indigenous people off-reserve, recent immigrants, single parents, and people with disabilities – are more likely to be in the lowest income groups.[21] For Indigenous people, the average income is about two-thirds that of non-Indigenous Canadians; Indigenous people are much more likely to be low-income earners; and Indigenous children are almost twice as likely to live in poverty. Immigrants also have about two-thirds the average income of non-immigrants and are more likely to be low-income earners. Racialized Canadians do slightly better, with an average income around 75% that of non-racialized Canadians.[22] Any policy that imposes added costs on lower-income groups will hit these groups particularly hard.

Carbon versus Non-carbon Workers. As we discussed in chapter 3, industries that use more carbon (either directly in their operations or indirectly such as through electricity) and that are more exposed to competition from other countries will tend to be hurt to a greater extent by the costs imposed by a carbon policy. We looked at various ways to mitigate these costs. "Cost to the industry" sounds neutral, but it means that those who work in those industries could lose their jobs. And a lot of Canadians work in these industries.

The oil and gas industry, for example, directly employs around 169,000 workers.[23] The Canadian Association of Petroleum Producers claims that number balloons to 528,000 jobs in Canada if you include not only those directly employed but also workers in companies supplying equipment, materials, and services.[24] The phase-out of coal would affect up to 4,000 workers in as many as fifty communities.[25]

The oil and gas sector was hit hard in 2020 by the oil price crash due to the jump in supply from Saudi Arabia and the drop in demand due to the pandemic.[26] Employment in the oil and gas sector fell by about 8% between March and May,[27] and the fallout might be even more widespread. As we discussed, the resource boom helped increase wages for low-skilled workers in Canada since 2000. The good times for a resource sector spill over to non-resource sectors within that same province: a resource boom leads to wage increases overall in resource-based provinces.[28] It can also have effects beyond the particular province itself. For example, a large number of workers living in Cape Breton commute to the oil sands for work. Such work not only provides those workers with wages, it may lead to increased average wages for workers in Cape Breton who do not commute, who could threaten to leave for work in the oil sands unless their wages are increased. Moreover, a resource boom would affect an even wider range of people and industries that depend on the incomes from these workers.

However we define the different sectors and the direct and indirect jobs, climate policy will have employment and wage effects. The wage effects may be partially progressive because those with higher incomes get a greater percentage of their income from employment.[29] However, wage cuts at the lower income levels can have significant effects on the well-being of those who are already struggling. There will also be trade-offs across industries. The BC carbon tax, for example, led to significant employment losses in carbon-intensive, trade-exposed industries (such as chemical manufacturing) but employment gains in some non-traded services (such as health care).[30] It also disproportionately hurt less-educated workers.[31] The bottom line is that carbon costs will hurt some workers more than others, and some workers potentially quite a lot.

Fairness and Climate Policy

This impact on current consumers and workers could be viewed as a short-term cost – that they are just the "losers" as Canada makes the necessary move to a more sustainable economy in the most efficient manner possible. An economist would see an efficient policy change, on one measure, as one where the "winners" gain more than the "losers" lose. The policy would be efficient if the winners could take their winnings and pay off all the losers' losses and still have something left over. Among the problems with this approach, of course, is that these payments are rarely made, so the losers tend to bear their losses.

A narrow use of this approach to climate policy asks if the reduction in the risk of global warming is worth the impact on GDP or on jobs.[32] The federal government's approach could at times be seen in this light. In 2016 it backed the Trans Mountain and Line 3 pipelines as well as a large LNG facility in BC. Increased emissions – from expanded oil sands or LNG production –were traded off against increased jobs and revenue. The government may have hoped for other benefits – that Albertans would accept its broader climate plan or that government revenues from the oil sales could be used for other public goods – but its decisions made the national emissions reductions targets harder to hit because of these increased emissions. Further, it created greater support for continued emissions by building up the oil sands or LNG companies and their workers: more money and more people are now invested in the continuation of these industries. The target is not only farther away, it is harder to reach.

We saw a slightly thicker vision of climate policy in chapter 3: we should try to reduce GHG emissions while growing the "green" economy. Jobs lost in fossil fuels or cement will be offset by jobs producing parts for wind mills or developing electric cars. If we grow the pie with green ingredients, no one has to lose.

However, not everyone may be able to share in the now-larger pie, as workers in the old economy might not be able to work in the green economy.

A still thicker approach would be based not only on economic growth but on what is fair. It has its roots in philosophy but connects to economics as well. One of the most well-known approaches to justice is what philosopher John Rawls called "justice as fairness."[33] Everyone should have "primary social goods" such as income, rights, and opportunities, and any inequalities are only justifiable if they work to the benefit of the least well-off in society (which is called the maximin principle).[34] We all want different things in life and should all have some basic ability to pursue those things.

Amartya Sen, a Nobel Prize–winning economist, built on this notion. Sen felt every individual needed to have the capability to "lead the kind of lives they value – and have reason to value."[35] A person's capabilities depends on her having certain "freedoms," including political freedom, economic facilities (such as the freedom to trade with others), social opportunities (such as education and health care), transparency guarantees (such as rights to disclosure of information), and protective security (protection of the vulnerable from such things as unemployment or poverty).

Sen argued that economic freedom is important but only one of these freedoms: "While economic prosperity helps people to have wider options and to lead more fulfilling lives, so do more education, better health care, finer medical attention, and other factors that causally influence the effective freedoms that people actually enjoy."[36] The goal of justice should be to try to equalize everyone's capabilities through these freedoms. This view would point towards a broader connection of climate action with other things we value, including health and education in addition to financial security.

The current term for climate policy built on notions of justice is a "just transition."[37] A just transition is not just about ensuring that we take climate action. We can move to a carbon-less economy in many ways: we could become green but increase the wealth of those at the top or we can move in a direction that decreases inequality. Canadian academic David Doorey notes that carbon policy "should include a theory of justice that recognizes that there will be costs and benefits to societies associated with climate change and the transition to a greener economy, which should be distributed in an equitable manner."[38]

We should aim generally, not only in the climate context, to move Canada towards greater justice. As Sen argues, this does not mean we need a consensus vision of what a fully just society would look like. Instead our policy decisions should always try to move us towards more rather than less justice.[39] It not only seems morally right but the most likely path to positive change. We will only get widespread support for action on climate change if we can show that any change is fair.[40] Fairness may not be sufficient to bring about action, but it is most likely necessary.

We may disagree on what fairness or justice would mean. For some, fairness prohibits visiting the costs of climate change only on those who are most vulnerable. Others may weigh innovation and long-term growth or lower costs more highly in order to ensure we have the resources to provide for future generations. What is important, though, is to take into account all the ways we impact people's lives.

Lost in Transition

This notion of fairness and a "just transition" is quite broad. To make it more concrete, we can think about what happens to two groups: vulnerable individuals and workers in carbon-intensive industries.

Vulnerable Individuals. Governments can, and generally do, have policies that deal with some of the obvious impacts of climate policy on vulnerable individuals. To take the clearest example, governments often funnel money from carbon pricing directly to those who may be hurt by the increased cost. To account for the potential effect of its carbon price backstop on families, the federal government returns the money from the tax to families. There is a general rebate to households plus a 10% extra top-up for households in small towns and rural areas.

The Parliamentary Budget Officer, an independent federal official, found that most families will actually get more from the federal payments than they pay out in the fuel levies.[41] The PBO estimated that the average household would pay $256 in the carbon price but would receive $300 back. These rebates reduce the negative effects in those provinces on which the federal government has imposed the carbon price (at the time Ontario, Manitoba, Saskatchewan, and Alberta).

BC's own carbon tax also tries to account for the differential impacts of the tax. The tax was initially designed to be revenue neutral – that is, the government would reduce other taxes to offset the increased revenue from the carbon tax. In 2012, for individuals, the impact of the tax was reduced in three main ways: a tax credit for low-income families, a cut of 5% in the tax rate in the two lowest tax brackets, and a lump-sum payment for northern and rural homeowners (tied to property values, income, and other factors). While the carbon tax itself hit lower-income groups less than upper-income, these measures further lessened the relative impact on low-income earners and rural and northern residents.[42]

Regulations may also be tailored to take account of their impacts on vulnerable groups almost if not equally as well as using revenue from carbon pricing.[43] Sometimes regulations cannot be explicitly tailored but may need to be tied to subsidies or other measures to reduce the impact. Ontario, for example, was a leader in Canada in getting rid of coal, closing its last coal-fired electricity generating plant in 2014 and shifting money towards renewables. The province, however, has struggled with the effects of the phase-out combined with other factors affecting electricity prices, particularly for those living in rural areas.[44]

As a further safety net, governments provide general transfers directly to vulnerable individuals such as low-income earners. These transfers have reduced inequality in Canada, with the most recent success story being the federal Canada Child Benefit, which has reduced child poverty. The federal government responded to the pandemic with broad temporary measures to help Canadians generally with the economic consequences of measures to address the public-health emergency. More generally still, we have health care and social programs that aim to help individuals in need.

Workers in Carbon-Intensive Industries. Governments are keenly aware of the impacts climate change policies have on vulnerable industries and on employment in these industries. As we saw in chapter 5, they have tried to ensure that industries that will bear the heaviest costs from climate policy get some kind of relief. The form of this relief for industry may not be optimal, but it is clear that concern about jobs (and revenue) is front and centre for our governments.

What about explicit help for those who lose their jobs? We have seen two main responses. First, the federal government and many provinces see developing new jobs in the "green" economy as key to the transition. The federal government's recent plan was, after all, called *A Healthy Environment and a Healthy Economy* and contained billions in support for creating a "clean industrial advantage" resting on the low-carbon economy.[45] It sought to "build back better" after the pandemic. Provinces also try these types of measures. Alberta, for example, is seeking ways to not only help find jobs in alternative forms of energy but also to develop and export technology to make oil and gas development cleaner. Subsidies, procurement rules, tax breaks, and local content rules are all ways governments have tried to get an advantage in the "green tech" market and find new sources of jobs. Canada might be well placed to take advantage of this shift, although both federal and provincial policies may well be insufficient at present.[46]

Second, Canada has both general and specific policies that aim to help those who have lost their jobs. While we can debate their sufficiency, Canadian tax and transfer programs support individuals who are unemployed or who are not earning much money.[47] Some are transfers with conditions and some are general retraining programs that individuals can access if they have lost their jobs or are underemployed. Health care and social programs offer another backstop.

More specifically, governments may aid individuals or communities harmed by the transition to a low-carbon economy. The federal government, for example, has created a program that provides support for communities that are or will be hurt by the phasing out of coal, although more is needed.[48] The federal government has more recently promised to put in place a *Just Transition Act* to address issues of workers in industries impacted by climate initiatives.[49] Its newest plan, however, is light on proposals to directly help workers, proposing funding for community infrastructure and a training benefit for all Canadians.[50]

Sharing the Risks

Canadian governments, then, are starting to take notice. They have policies that reduce emissions somewhat as well as aim to offset the short-term costs. They are trying to keep fossil fuel industries in place to preserve jobs and government revenue and have taken some steps to reduce the effects of the emissions reductions on more vulnerable individuals. These policies are relatively easy to adopt when the emissions reductions are not large. We can encompass the transfers to vulnerable individuals within our current conception of the economy: we keep the economy growing to the maximum extent we can (subject to moderate limits on emissions) and distribute the gains to the most vulnerable. Governments have also begun to think about ways to prepare for a shift away from fossil fuels, which it is increasingly clear is coming. All of this helps, but it leaves much of the concern in place.

Canada is at a critical juncture. The policies listed above will not be sufficient to bring about the change needed. Economic growth and markets will be an important part of the long-run solution. Many people wrote about how to use the "opportunity" of the pandemic crisis to move towards a net-zero economy.[51] It is true that there is a lot of money in play; both federal and provincial governments want to get money to industries to keep them afloat. Channelling this money to shift towards a low-carbon or no-carbon economy may help find solutions and build support for change from companies and different sectors.[52] The chance to leverage this type of spending may not come again for a long time, given the hole that governments are in after passing out this money. The move could include those in the fossil fuel industry who are pushing for development of new ways of producing energy. It is worth remembering that for ozone depletion, global action was successful but it took not only a concern for the environment but also an industry that stood to gain and therefore was willing to push for change.[53]

However, trying to grow our way out of trouble will not work on its own. Workers and others are less likely to trust the change of direction and more likely to fear a loss without a subsequent gain – that they will not share in the new growth – unless their interests are more directly considered. It is hard to support something you feel is going to harm you, your family, and your community. The Green New Deal in the United States and its Canadian counterpart (and precursor) Leap attempt to deal with this issue.[54] They call for a wide-ranging move towards social justice, tying together climate change, inequality, discrimination, and Indigenous rights.

These calls for combining direct climate goals with other social aims have traction around the world. The Biden administration in the United States has adopted elements of the Green New Deal and proposes to invest billions in a shift to a lower-carbon economy as well as to address inequality and discrimination. In the European Union, a proposal led by economist Thomas Piketty advocated addressing climate change and inequality, including through carbon taxes but also other taxes on corporations and the wealthy.[55] The EU subsequently adopted the European Green Deal Investment Plan that involves both public investment and incentives for private investment in a transition to a carbon-free economy as well as a "just transition mechanism" aimed at providing funding for workers and communities most hurt by the transition.[56] At the UN, there are calls for a "just transition" that recognizes the risk that workers in certain industries may be left behind in the drive towards decarbonization.[57]

The US Green New Deal as a whole is not a perfect fit in the Canadian context. The spirit of many of the proposals is relevant to Canada, such as the drive to reduce emissions and deal with inequality both in terms of income and regarding visible minorities and Indigenous peoples. On the other hand, some of the proposals already exist here, such as universal health care. We need to consider the Green New Deal's applicability within the Canadian context, including Canada's federal structure and policy history.

Moreover, there are debates about how useful this broad approach is to immediate action on climate change. For example, economist Mark Jaccard contends that it makes fighting climate change even more difficult by trying to graft on other agendas. He argues that we cannot get sidetracked "from our essential task of quickly decarbonizing critical sectors of the energy system," for which we do not have to get rid of capitalism.[58] Otherwise we make the problem harder when in fact, for Jaccard, we have the basic tools we need to move off coal and gasoline. It is a difficult choice. Even without going as far as Leap or the Green New Deal's aspirations, a broad reform agenda risks losing time and alienating people potentially on board with reducing emissions but against wholesale changes to society. It also may dilute our ability to monitor how discretion is used to fight climate change. To the extent regulators are pursuing multiple goals, it is harder to hold them accountable to not meeting emissions reductions targets (they can point to trade-offs with other aims to justify not reducing emissions).

Canada's history points towards a middle ground, one that draws on its policies spreading risks. Over the course of the twentieth century, both the federal and provincial governments began managing more and more risks that had previously been left to the market: unemployment, sickness, aging, the environment, and occupational health and safety, along with risks to businesses.[59] Canada has gone through three broad stages in the growth of government's role in managing risks. The first stage, prior to the Second World War, saw government leave much of this risk to the market. Its main role was promoting economic growth by reducing risks for business.

The Depression ushered in the second stage with governments, especially the federal government, taking on risks to individuals through programs relating to unemployment insurance, pensions, and poverty. Such risks were seen as salient to a large swath of the public and, following on influential reports from the Rowell-Sirois Commission in 1940 and the Marsh Report on Social Security for Canada in 1943, the government responded to demands for more protection. As we will see, this move also had implications for the balance of powers between the federal and provincial governments. This period lasted into the early 1970s, with governments beginning to take on an increasing array of risks, including to the environment, health and safety, and human rights.

A third stage was ushered in during the 1970s with governments responding to concerns over budget deficits by reducing their role in managing risks. This period saw cuts to government programs in areas like health and education but also to transfers to individuals like welfare. One partial explanation may be that the risks were not as salient to middle-and-upper-income Canadians as they had been following the Depression, leading to less political support. Governments in the 1980s and 1990s also began deregulating in some areas, seeking to rationalize the regulatory "burden" on industry. This period saw the rise in inequality we discussed earlier in this chapter.

Canada has, then, shifted over time from leaving risks to markets to governments taking on a much stronger role in sharing risks across Canadians and then back to more of an emphasis on markets. Recent crises have reignited recognition of government's role as risk manager, including the financial crisis of 2008 and, most recently, the pandemic. The federal and provincial governments responded by developing programs to reduce the harms individuals suffered from unemployment, poverty, or illness.

The difference with the climate crisis is that the risks are longer term and more slowly developing. Government's main response has been to try to foster new industries to create jobs in the long term. The risks from unemployment and to individuals and communities are largely left to the market to sort out. It is this friction in the transition that can give rise to fear and, potentially, a lack of support. Governments have spoken of a "just transition" but have done little to actually deal with the costs of this transition to individuals. They have emphasized jobs in the new "green" economy without figuring out whether the new jobs will be as secure or well paying as prior work.[60] We need to address a range of these risks, not only because it is the right thing to do but because Canada will not achieve effective near-term action without doing so. Luckily, addressing these risks fits with Canada's history of how governments have dealt with such risks in the past.

The difficulty lies in balancing the virtues of markets with the government's ability to spread or share risks. Both economic stability and finding solutions to climate change depend on finding the right mix of markets and governments. Many good things have come from markets and the desire to innovate. However, greener growth will only be a fair, achievable path forward if it makes people better off. The worry is that the growth will actually benefit the wealthiest people, disadvantage the most vulnerable, and exacerbate rather than alleviate poverty. The pandemic showed how reducing risks can lead to unfair distributions of risks and gains.

As we have seen, dealing with the extra costs of climate policy on low-income earners seems straightforward. If the costs of these measures are falling disproportionately on lower income groups or Indigenous peoples or people in rural areas, Canada has experience with tax and transfer systems that can help alleviate these costs. Canadian governments have used the transfer system to good effect with some carbon pricing systems, for example. Note that we are talking here about largely about the economic costs, as opposed to broader adaptation costs.

The more difficult issue is what to do about those whose jobs will be lost in the transition. Some will face costs of climate action that other Canadians do not for no other reason than that they work in an industry that is now disfavoured. Canada, again, has some experience with such transitions and can learn from past failures and successes. The attempt to deal with the transition of workers and communities away from reliance on cod fishing comes immediately to mind.

Canada's current suite of fixes can make some difference – transfers, retraining, subsidies for new industries. Moreover, in its 2021 budget the federal government announced $1.7 billion for retraining workers, including $55 million for transitioning related to decarbonization. These solutions are not without their potential problems, however. There is concern that such transition policies help some workers (generally white and male) get support and new jobs but do not help marginalized workers (and particularly those in jobs that support but are not directly related to the coal industry). Transition policies should obviously also help these marginalized workers.[61]

However, more is needed. Many workers in fossil fuel industries may see their jobs and their communities as central to their identity. Economists at times seem to view the problem as helping to retrain workers who can then move to wherever there are jobs. There is a cost, though, beyond just the lost job when workers, such as those in the coal industry, identify closely with where they live and work. Any just transition policy must take account of these connections to identity, including through attempts to create local jobs, engage local communities, and invest in local infrastructure. [62]

We also need to think about people who are not yet in these industries and what it means to prepare them for a very different economy. "Human capital" tools like providing education are not overly effective at reducing income inequality in the short run. Some benefits arise through increasing subsidies for lower-income students in universities and colleges, but the big benefits come from better early childhood education.[63] Something analogous seems true for this transition. Retraining will be part of the solution, but big gains will come from ensuring an education system that opens up opportunities for all Canadians.

These policies need to be constructed in a manner that does not undermine creativity or innovation in the economy, such as by creating perverse incentives to those receiving aid. However, such incentive effects have been central to Canadian discussions of managing social risks since at least the 1960s. Our governments have experience designing these programs. We just need to recognize the gap in risk management that governments are creating by leaving the transition largely to the market.

Finding the Bootleggers

We need a stronger bootlegger group – a strong voice for change from those who would benefit. It can seem at times like we have a form of myopia – we see the immediate self-interest from continuing emissions but are wilfully blind to the long-term costs.[64] We need to act to deal with these long-term harms. Regardless, change is coming as countries start to take significant action on climate change, and the policy space in Canada to avoid this transition is rapidly shrinking. We need people to appreciate that climate action will be in their best interests, and this means thinking about what their best interests actually are as well as understanding their concerns. We need to understand why some people are resistant to something that to others seems so blindingly obvious.

Both fairness and pragmatism point to recognizing and dealing with the fears that underlie the shift.[65] We cannot rest on the hope that growth will take care of everyone. If we care about those who are vulnerable in this time of change, we need to think carefully and empathetically about how they will be affected. We have to build on Canada's history of risk-sharing to be able to take the steps necessary to fairly and effectively address emissions. Such policies will require federal and provincial cooperation. The federal government has some levers but the provinces hold many of the powers that will be important – over education, health, and social policies. Coordination and cooperation problems across levels of government become bigger and more complex when we have to think about not only how to significantly reduce emissions but how to defray the costs. We turn to these issues next.

chapter eight

Strengthening the National Community

In the previous chapter we considered how we may be able to reduce individual opposition to change if we think about what those individuals need and address their fears of the transition. A related concern is that climate policies that impose huge costs on the fossil fuel industry in Canada or harm emissions-intensive industries will destroy the source of Canada's prosperity and growth. Canada has relied heavily on these industries to not only provide jobs and income but support the social system Canadians are justly proud of – the health care and education systems and social welfare programs. The worry is that if our governments take strong action on emissions or do not give these industries support in dealing with issues such as the pandemic, they will not have the money to pay for these programs. Economic growth will fall off and, with it, revenues from income and from corporate taxes or royalties; the result will be that the education system worsens, the health of Canadians declines, and inequality increases. Economist Mark Jaccard calls this a "Faustian pact" – we accept the environmental harm because it supports our lifestyles.[1]

Certainly, resources have driven much of Canadian growth, and we need to take the loss of government revenue very seriously. However, in chapter 7 we discussed Sen's theory of what a fair society would look like – people should be given the tools to attempt to attain what they value in life: political freedoms, economic freedom, social opportunities, transparency guarantees, and protective security, which includes protection against such ills as unemployment or poverty. Building such a society requires resources, but Sen does not claim that a country needs to obtain economic security first in order to be able to afford the other freedoms. They are all interwoven and depend on each other. These freedoms are both the goal of a fair society and the means for achieving it.

How bad would it be if Canadian governments imposed large costs through climate policies? The first step is to try to figure out how much revenue fossil-fuel-based industries, which are seen as an important driver of economic growth, provide to support government services.

Filling the Public Purse

Resource industries like oil and gas provide governments with direct revenue by, for example, paying for the right to develop an oil deposit and paying a "royalty" for the amount of oil they actually end up producing. The provinces in particular benefit from such revenue, since they have control over natural resources under the Constitution, creating disparities in regional wealth. Governments also receive corporate and personal income taxes.

Between 2009 and 2013, all Canadian governments gathered $79 billion in revenues from all non-renewable resources (such as through royalties).[2] Government revenues from oil and gas averaged about $15 billion per year from 2015–19, and for 2013–17 the energy sector contributed about 8% of taxes paid by all industries.[3] Four provinces – Alberta, BC, Saskatchewan, and Newfoundland and Labrador – received 90% of this revenue, with Alberta getting the lion's share at about $46 billion over the period, Newfoundland and Labrador $13 billion, Saskatchewan $10 billion, and BC $4 billion. The rest of Canada combined received about $4 billion.

For some provinces the resource revenue made up a significant share of total revenues; for 2009–14, taxes on non-renewable resources made up about 22% of total revenues for Alberta, 17.5% for Saskatchewan, and about 33% for Newfoundland and Labrador. However, this reliance has brought concerns. Economist Trevor Tombe, for example, has noted that "Alberta is addicted to resource revenues."[4] The share of revenue the Alberta government received from non-renewable resources jumped tremendously starting in the 1950s and has been as much as 60% of government revenue, though it is extremely volatile, dropping to below 20% in some periods. This revenue has allowed Alberta to grow its public spending to levels consistently higher than the ten-province average over the past twenty years, despite the province's having generally lower tax rates.[5] In 2020 a drop in oil prices resulted in falling revenues, and the province struggled with deficits, causing it to significantly cut a wide range of provincial services.[6] More recently, oil prices have rebounded, improving the fiscal situation at least in the near term.[7]

The federal government also benefits from resource industries. It directly receives revenue from royalties such as for off-shore drilling, but more important is the money it receives from corporate and personal income taxes. It is difficult to find precise figures for federal revenues from the oil and gas sector. According to Tombe, Alberta accounts for almost one-fifth of the taxable income of individuals and corporations.[8] The federal government received about $2.7 billion from the oil and gas industry in corporate taxes in 2009–10.[9] Given that the federal government's expenditures in 2018–19 were about $346 billion, such revenue, even if it has increased, is important but not overwhelmingly large.

Allowing continued emissions could bring another big payout for governments. They can continue to collect revenues from any carbon price, provided the price is not so high as to actually eliminate emissions. The BC carbon tax will bring in about $1.6 billion in 2019–20.[10] The PBO estimated that the $100 per tonne federal carbon price necessary to achieve the Paris targets would have generated about $53 billion in tax revenue had it been implemented in 2013.[11] In Ontario, the government generated about $2.9 billion in revenue from its auctions under its cap-and-trade program before they were cancelled, and because of the cancelation will miss out on about $1.9 billion per year for 2019–22.[12] Given the differences between provinces in natural resources and emissions, there will be significant variance across provinces in the revenue from a carbon price.[13] However, often carbon prices are intended to be revenue neutral. The federal government, for example, returns much of the money from its backstop carbon price to households.

On the other hand, leaving aside the human and environmental toll in other countries, failing to act on climate change will bring its own massive costs to Canadian governments. As we discussed in chapter 2, Canada will face more extreme heat and extreme cold, more precipitation but shortages of fresh water in summer, floods, drought, wildfire, and so on. The heat waves and floods in western Canada in 2021 provide ample and vivid illustration of these potential impacts. While scientists are making strides in determining which increases to attribute to climate change,[14] it is clear governments will incur significant costs for dealing with issues such as floods and wildfires. The Canadian Institute for Climate Choices calls these more obvious costs "the tip of the iceberg" as there are so many more, hidden costs.[15] Costs also potentially arise from addressing health effects such as infectious diseases as well as respiratory and cardiovascular disease from adapting to climate change.[16] Those who are most vulnerable – the North and vulnerable populations in other regions – are most likely to bear these costs.

Estimating the costs of non-action is made difficult by the fact that the federal and most provincial governments have failed to assess the risks of climate change or set out comprehensive adaptation plans.[17] According to the Canadian Institute for Climate Choices, extreme weather events are happening more often and becoming more costly. They estimate the losses rose "from an average of $8.3 million per event in the 1970s to an average $112 million between 2010 and 2019, including public and private costs."[18]

In terms of direct government costs, the federal government runs the Disaster Financial Assistance Arrangements program, which reimburses provinces and individuals for natural or human-made disasters. It shares an increasing proportion as the costs go up, rising to 90%.[19] The Parliamentary Budget Officer noted that the payments would be $580 million in 2017–18 and, going forward, the payouts will increase to about $900 million per year. [20] The Federation of Canadian Municipalities and the Insurance Bureau of Canada put the cost of municipal infrastructure and local adaptation measures to prepare for the impacts of climate change at $5.3 billion per year (about 0.26% of national GDP).[21]

The further costs are potentially massive and incalculable. For example, the number of premature deaths in Canada each year from air pollution due to fossil fuel combustion is estimated to be 14,400, and the overall health-related costs of fine-particulate air pollution is about $53.5 billion.[22] How do we count the costs to Canada of mass global migration? Of the loss of species and ecosystems? These questions are even more difficult to answer. The Canadian Institute for Climate Choices is trying to estimate climate costs. But we need to take those costs seriously if we are debating whether addressing emissions is worth the cost.

Subsidies

It is clear that the fossil fuel industry, and industries that are emissions intensive as a whole, are important to the Canadian economy but also important in helping to fund services that are central to ensuring fairness. However, in order to keep these revenues coming in and maintain growth and jobs, the federal and provincial governments have given money directly and indirectly to these industries. Such subsidies muddy this story of government reliance on fossil fuel revenues.

In addition to the direct costs of these subsidies, economists always worry that government subsidies either give industry money they do not need or allow industries that are not profitable to survive. Offering tax advantages to the oil and gas industry may induce people to invest in that sector when they would not otherwise do so.[23] And fossil fuels already get a significant subsidy – emissions into the atmosphere have essentially been free, a hidden subsidy underlying production and use of these fuels.

In 2009 the G20 group of countries, which includes Canada, committed to "rationalize and phase out over the medium-term inefficient fossil fuel subsidies that encourage wasteful consumption."[24] However, Canadian governments were estimated to give over $3 billion in annual subsidies to the industry in 2015[25] and possibly about $5 billion in 2015 and 2016.[26] Canada provided the most government support to oil and gas per unit of GDP of any G7 country.[27]

In the 2015 federal election campaign, Justin Trudeau promised to phase these subsidies out,[28] and in 2016 Canada, the United States, and Mexico committed to eliminating such subsidies by 2025.[29] Yet by one estimate, the Trudeau government provided about $1.6 billion in production subsidies to the fossil fuel industry in 2018.[30] Further, the federal government has invested in infrastructure that helps various industries, most notably through its 2018 purchase of the Trans Mountain Pipeline, which may cost upwards of $12.6 billion.[31] More recently, it has provided support for these industries, among others, during the pandemic.[32] Canadian governments (both federal and provincial) have provided about $23 billion in subsidies to support pipelines since 2018, according to the International Institute on Sustainable Development.[33]

Federal subsidies fell to about $600 million in 2019.[34] However, while the Department of Finance Canada and Environment and Climate Change Canada (ECCC) claims to be phasing out "inefficient" tax subsidies, it is difficult to determine the full extent of such subsidies. The federal Environment Commissioner has criticized the definition of "inefficient" subsidies as being overly broad and vague and charged that the federal government fails to sufficiently account for economic, social, and environmental sustainability.[35] Moreover, subsidies rose again in 2020, more than tripling to about $1.9 billion, in part due to COVID-related funding, with Environmental Defence putting the number even higher at $3.2 billion in direct aid.[36] These subsidies can include money to reduce emissions, such as the recent subsidies for carbon capture and storage.

The provinces are also in this game. Environmental Defence and the International Institute for Sustainable Development reported that Alberta in 2018 gave about $2 billion in support to its fossil fuel industry, including in royalty reductions and tax exemptions, although the province only got back about $2.4 billion in royalties.[37] Ontario provided almost $700 million in subsidies for production of fossil fuels in 2018, including tax breaks for fuels used in air, rail, and agriculture and a natural gas expansion program.[38]

BC gave over $800 million in 2017–18, including to the LNG Canada project.[39]

Similar moves are occurring elsewhere. The UN estimates most of the major fossil-fuel-producing countries are not going to meet their Paris commitments, partly because many are intent on increasing production of fossil fuels.[40] It states that, in terms of just direct subsidies, governments gave over $24 billion in 2017. Another estimate saw G20 countries providing more than $3 trillion in subsidies for fossil fuels since the signing of the Paris Agreement.[41] The United States alone gave about $8.2 billion (USD) of subsidies in 2016, with one study finding that about half of all new US investments in oil require government subsidies to be profitable.[42] Such subsidies are another form of collective action problem: all countries may prefer not to give any subsidies but feel the need to if others do.

The fossil fuel industry is clearly a significant employer and has been central to the finances of some provinces. Subsidies to the fossil fuel industry complicate this picture. Fossil fuels got a mention in the federal government's climate plan, which noted it will "deliver on Canada's G20 commitment to phase-out all inefficient fossil fuel subsidies by 2025."[43] More recently, the Liberal party promised during the 2021 election campaign to eliminate fossil fuel subsidies by 2023. The task has proven very difficult so far.

Equalization and Stability

Fossil fuels are thus important to government provision of public services, though less so than it might seem. Developing support for strong climate action as well as maintaining a commitment to fairness will require us to address concerns about loss of such services, particularly for fossil-fuel-rich provinces and individuals in those provinces. Fortunately, Canada has a framework that can help – its equalization and stabilization programs. Both are important parts of the economic framework of Canada, and neither has an immediately obvious connection to our progress on climate change. Yet both will likely be key to our thinking about possible solutions for the climate crisis.

We should start by being clear about the difference between these two programs. Equalization has a long and contentious history in Canada.[44] Section 36 of the Constitution states that the federal government is "committed to the principle of making equalization payments to ensure that provincial governments have sufficient revenues to provide reasonably comparable levels of public services at reasonably comparable levels of taxation." The federal government's equalization program addresses this requirement. The fiscal stabilization program, on the other hand, is a smaller program aimed at helping provinces that have been hurt by an economic downturn.

Under the equalization program, the federal government collects money through taxes and then distributes money to provinces less able to raise revenue to pay for government programs like education. The federal government determines who will receive funds by figuring out the "fiscal capacity" of each province – essentially the amount of revenue each province could raise if they had similar taxes. Provinces with strong economies will be able to raise more than those with weak economies, and the equalization program is meant to mitigate the imbalances this creates. [45]

In terms of public perception, the equalization program divides the country into "have" and "have-not" provinces, with the have-not provinces being the ones getting equalization payments. It is not a clear West vs. East issue: Manitoba receives payments and Saskatchewan has in the past. In fact, all provinces have at one time or other received payments, although Alberta has not been on the receiving end since the 1960s. Ontario became a have-not province in the financial crisis of 2008. Quebec, which gets the bulk of the equalization payments, was set to receive $13 billion in 2020–21.[46] Currently five provinces receive payments: Manitoba, New Brunswick, Nova Scotia, Prince Edward Island, and Quebec. Alberta is still a have province, even during its recent struggles with low oil prices.

One of the most troublesome issues is whether revenue from natural resources should be included in equalization. While the provinces own the resources under the Constitution, resource booms cause difficulties for equalization. According to economist Robin Boadway, a massive oil boom in one region would cause problems even for a unitary country but is much more difficult to manage for a federation like Canada. [47] One problem is the so-called Dutch Disease where a boom in a resource industry hurts the country's manufacturing sector through avenues such as an increase in exchange rates (which harms exports of manufactured goods). A country, unitary or federal, would also have to decide how much to invest in social programs across the country, diversifying the economy, or building infrastructure to serve that industry (such as a pipeline).

In a federal system, there is the added worry about how one province's revenues from a resource boom impact other provinces. Workers may, for example, be drawn into the booming province, seeking more public services at lower taxes. The resource-rich province could also use its revenues to entice industry away from other provinces that lack these resources.

Resource revenues in one region thus create issues both of equity (in the sense of ensuring all Canadians have potential access to the same level of public services) and efficiency (in its potential to create incentives for individuals or industry to move to that province). The current equalization formula accounts for the special place of resource revenues, but in a complicated way. The formula only includes 50% of a province's revenues from resources or none, whichever is to the advantage of the province. In part, this rule provides an incentive for receiving provinces to develop those resources.[48]

As a result, the program attempts to fulfil the constitutional commitment to equalize potential access to services and reduce adverse incentives from resource booms but at the same time to respect provincial control of resource revenues. Provinces like Alberta, however, do not like the idea that they are subsidizing other provinces.[49] Although the have provinces do not send money to the have-not provinces – payments come from the federal government's tax revenue –concerns over inequity are important when thinking about the political nature of equalization. The

Alberta government recently launched a Fair Deal Panel to look into how the province fared relative to other provinces,[50] and also held a referendum in the fall of 2021 in which a majority of Albertans supported removing equalization from the Constitution.[51]

The other program, the fiscal stabilization program, works a little differently.[52] It is designed to provide funds from the federal government should a province experience a sudden and substantial drop in revenue. Because a province's resource revenues can swing so much relative to revenues from other industries, there is a different threshold for getting help depending on what is going wrong. In general, a province can get stabilization money if its non-resource revenue falls by over 5 percent and if its resource revenue falls by over 50 percent.[53] There is less money available under this program; it was capped at $60 per person in 1987.

Economist Bev Dahlby suggests this program should function more explicitly as insurance for the provinces. He notes that, while there are other reasons, "a fiscal-insurance program can be one of the motivations for a rich but risky region to belong to a federation."[54] According to Dahlby, the payments to the federation should be seen as a "fiscal-insurance premium," paid out if things go bad through the stabilization program. Like insurance, the program should cover large, unpredicted losses in all revenues but include a form of deductible (an amount the province would have to cover itself) to ensure provinces maintain their own revenue-stabilization funds to offset short- to long-term declines.[55]

In a way, that is how the stabilization program works, just on a meagre basis. The program has its roots in the post-war period but its conditions have changed over time. Now, with the $60 per person cap and the requirement that resource revenue drop by more than 50%, Tombe argues it "has atrophied to near insignificance."[56] He similarly suggests revisiting the need for the cap on payments as well as whether the program should include resource revenue.[57] The difficulty, for Tombe, is that including resource revenue (even with the threshold) means that provinces with abundant natural resources do not have to be as responsible with their budgets. He specifically points to Alberta as needing to find other, more stable forms of government revenue (such as a sales tax) rather than hoping for stabilization funds.

One further point about transfers. Other federal transfers such as the Canada Health Transfer and the Canada Social Transfer as well as federal government spending and taxes are important to equalizing public services across the country.[58] In fact, federal redistribution from programs such as the Canada Health Transfer is much larger than the equalization payments – making up about two-thirds of the federal flow of money as opposed to about one-third from programs like equalization.[59]

The two main net contributors to federal revenues (that is, payments into federal revenue minus receipts of federal transfers) since the 1960s have been Ontario and Alberta, with Ontario having the higher total contribution over the period but Alberta a much higher per capita contribution, particularly in recent years.[60] Tombe notes that, given the same tax rates but different tax bases, provinces with higher-than-average income (again, like Alberta) send more to Ottawa in the way of personal income tax. On the plus side, this means that workers in these provinces earn more.[61]

As Tombe has noted, "federal transfers are essential to Canada's fiscal landscape, and have been since Confederation, but achieving stable, equitable, and efficient arrangements is difficult."[62] At less than 2%, the overall size of these transfers in Canada is smaller as a percentage of GDP than at any time since the 1960s. The stabilization program in particular is failing to help resource-dependent provinces.[63] However, transfers provide a solid foundation of sharing within Canada.

Sharing the Risk

Underlying much of what we have discussed in this chapter is not so much the money spent as it is the story we tell ourselves about it. A good deal of the recent discussion of this system has been negative and divisive. We need to take a step back and think about how Canada works. Canada is a federation organized to ensure the spreading of the risks and benefits of life. This is baked into the Constitution not only directly in the requirement for equalization but also in what the federal and provincial governments can do. The fact that the provinces control matters such as health care, welfare, and education and also have the benefit of transfers from the federal government shows a commitment to ensuring all Canadians have the basis for living the lives they have reason to value, as Sen would say. There are some glaring gaps – the treatment of Indigenous peoples and the partial though perhaps improving attempts to deal with poverty come to mind – but fundamentally, Canada is built to share risks and benefits.

Back in 2006, economist Robin Boadway asked, "Do we define our sharing community primarily at the national level or at the provincial level?"[64] As with many things, Canada has tried to do a bit of both. The Constitution gives power over resource revenues mostly to the provinces. Equalization provides the basis for public services for all Canadians, but with an adjustment to recognize this ownership of resources. Other federal transfers to provinces help them fulfil their responsibilities. The stabilization program helps those provinces hit by short-term troubles. These programs underlie the notion that "an important element of the benefits to a region from being part of a federation is the gains from risk sharing."[65]

Life comes in cycles – as Alberta clearly shows. Just in recent years, it had a tremendous boom followed by a recession, and then in 2019 the Conference Board of Canada predicted that Alberta would be the second-fastest-growing province economically in 2020.[66] However, the dual shocks of Saudi Arabia flooding the market with oil and the pandemic killed demand, hammering the Alberta economy. Tombe points to the federal government's response to the pandemic as "demonstrating the value of a broad federation that pools risk with measures like income support to individuals and businesses, transfers to provincial governments, and more."[67]

What does all this mean for climate policy? Canada faces a radical change if it is to become net-zero by 2050. We can argue forever about whether Alberta and the other oil-producing provinces should have been thinking about the transition earlier or whether provinces receiving equalization payments are hypocrites for accepting payments at least implicitly founded on developing fossil fuels. It gets us nowhere, however. Meanwhile, the clock is ticking.

To be true to the spirit of a national sharing community, the costs of change must be shared. Fortunately, Canada can build from a strong federal base to make that happen. More money needs to be channelled into helping make the move away from both consuming and producing fossil fuels. As a country, Canada consciously adopted policies to make it cheaper and easier to develop fossil fuels. Industries were born, jobs created, governments supported. We benefited as a nation; now, as a nation, we need to pay to change.

We have discussed government funding of emissions reductions, which the federal government appears anxious to do. We are concerned here with a different, though related, issue: how to address the loss of government revenue from strong climate action. We need to think about how to deal with this loss of revenue to the provinces and the associated incentive to delay action.

Equalization, stabilization, transfers are not the problem; they are the solution. Canadian governments need to build from both the spirit and form of current risk-sharing programs. Some changes could be made within the existing system. Equalization, for example, could be more explicitly connected to the fight against climate change. While many of the large emitting provinces tend to not be on the receiving end, the equalization program could take into account revenue from carbon prices to provide incentives to the receiving provinces to reduce emissions.[68] For stabilization, a (very controversial) possibility is to not cover losses in government revenue related to fossil fuel production in order to force provinces to find more stable sources of revenue and rely less on fossil fuels.[69] The difficulty is that we know that climate action is going to reduce these revenues, and it may seem punitive to knowingly reduce the revenue without any compensation.

More broadly, the motivation behind the stabilization program could provide a foundation to mitigate the downsides of the transition to a low- or no-carbon economy. Equalization was set up to deal with persistent differences across provinces, while stabilization is meant to address short-term, exogenous shocks – that is, events outside the control of the province.[70] Climate change has elements of both. Addressing climate change will lead to a long-term loss of revenue for many resource provinces. At the same time, it will involve a large short-term cost that is at least partially within their control.

One possibility is for the federal governments to tie payments to how much fossil-fuel-producing provinces like Alberta lose in revenue from strong climate action. Sharing some of those losses would recognize that such provinces are being asked to bear the cost of cutting back something from which all Canadians bene-fited, due to a crisis that is at least partially outside their control. It is in a way a stabilization program, as payments offset transitional, short-term losses. Tying payments to emissions reductions reduces the incentive for provinces to keep production high. Alternatively, conditions could be placed on the use of the money – for diversi-fying the economy, for instance – although that goes against how agnostic the federal government has generally been on how prov-inces use their revenues.[71]

A simpler route would be for the federal government to give block grants to provinces like it does with the Canada Health Transfer. Seth Klein, for example, argues for a Climate Emergency Just Transitions Transfer.[72] The money could be tied to conditions on reducing emissions or diversifying the economy. Unfortunately, the provinces are suspicious of block grants because the federal government has a history of reducing them. The Liberal govern-ment has promised a form of such funding by proposing a $2 bil-lion "Futures Fund" to aid in the "economic diversification" in Alberta, Saskatchewan, and Newfoundland and Labrador.

One last question: if a province like Alberta receives money to diversify its economy, should they pay it back if they are success-ful? The payments could be seen as an investment, perhaps mak-ing the program politically more palatable in the non-receiving provinces. However, in one sense it will come back anyway – to the extent the receiving provinces are successful in growing a new economy, the money comes back to everyone through equalization payments. Moreover, as all Canadians gained from promoting the fossil fuel sector, all should share in the costs of change. It is not a "loan" in that sense – it is a recognition that we all need to work together to address a problem we all created.

Some of the money for such payments could be found by ending subsidies for the production of fossil fuels. This money could be channelled to the provincial governments as opposed to industry. Industries could still receive subsides for innovation in reducing emissions or, even more importantly, for developing other forms of (renewable) energy. Such subsidies are central to the newest federal climate plan. The difficult part is ensuring these funds are not just replacing funds that would have been spent anyway.

Tombe has argued that Albertans "historically relied on oil revenue to fund profligate spending; we can do so no longer."[73] This same statement holds for the country as a whole. Canada has relied on a way of living and, in part, on an industry that we now know harms the environment. We can no longer afford to do so. We need to share the burden together.

chapter nine

Cultivating Cooperation

In his book *Exit, Voice and Loyalty,* political economist Albert O. Hirschman wrote about what happens to organizations – firms, voluntary groups, governments – when things start to go wrong.[1] He discussed two main mechanisms for change: exit and voice. Members can leave the organization (exit), which puts pressure on it to adjust, or they can stay in the organization and try to use political mechanisms (voice) to bring about change.

If a parent is unhappy with her daughter's school, for example, she can move her to another school. The school gets information that some people are dissatisfied with the way it is being run, but possibly not why. Also, the school loses the resources and any time or resources the parent would have invested in trying to make the school better. If the parent keeps her daughter in the school, she will need to engage in the messy business of exercising her voice, arguing with the school about the best way to educate the students. A Canadian political example is the Quebec referendum on leaving the country, as opposed to remaining in the federation and seeking to alter policies in Parliament.

Canada is in a similar situation in trying to reduce national emissions. One view is that the federal government should take strong action to move Canada towards meeting its targets through imposing a plan on the provinces. The scope of the problem is global and provinces, on this view, are too focused on local issues to take sufficiently strong action. The provinces can use their voice to have input into the national policy, but all need to stay in the plan.

Others, however, worry the federal government is not attuned to what is going on in the individual provinces. Some Albertans, for example, feel there is a lack of understanding or caring about the negative impact strong climate action will have on industries that are important to their provincial economy. The Liberal government's approval, and subsequent purchase, of the Trans Mountain pipeline was seen as insincere and insufficient. After the Liberal electoral victory in 2019, some in the West even floated the idea of a "Wexit," the West taking their resources and leaving the country. However, the less extreme version is what we saw in chapter 5 – provinces going their own way on climate policy.

The issue of the balance of power between the federal and provincial governments has troubled Canada since Confederation and is at the heart of ongoing debates about climate change. It has sparked elections, court battles, and even a referendum. This chapter looks at whether the Canadian federal structure is helpful or harmful to the fight against climate change. To start, we need to think about who really is best to lead on climate action: the federal government or the provinces?

Act Globally – Act Locally

It is reasonable to argue that since climate change is a global issue, the federal government should ultimately take responsibility on how Canada addresses it. After all, issues like national defence and international trade fall to the federal government. And, as we have seen, some provinces are less willing than others to reduce emissions. The federal government, on this view, can transcend provincial reluctance and take account of a broader range of interests. It may be less susceptible to the interests and advocacy of local industry. Federal government regulation helps avoid the "race to the bottom" problem – provincial governments competing to have the weakest environmental regulations in order to attract or keep industry.

National regulation is not perfect. The federal government may be less sensitive to differences in how regulations impact different provinces. Decision-making may be better placed at a lower level if the more local government is better aware of and more responsive to the voices of people who are adversely affected. The Supreme Court of Canada has recognized the idea (called "subsidiarity") that "law-making and implementation are often best achieved at a level of government that is not only effective, but also closest to the citizens affected and thus most responsive to their needs, to local distinctiveness, and to population diversity."[2]

Such local tailoring means emissions reductions may be undertaken at lower cost and may be more attuned to local benefits to the region.[3] For example, the Alberta government may have better knowledge than the federal government of the options for reducing its use of coal for electricity generation, but also may be more keenly aware of any added benefits such as improved air quality. Albertan regulation would then be more efficient, and these local, added benefits may increase the willingness of Albertans to bear the cost of reducing coal use.

Economists argue it is efficient for the provinces to have some powers. If people reasonably differ on the appropriate policy or group of policies (and can move easily), we may want to give them the choice of living in provinces that align with those preferences. People who do not like New Brunswick's policies on education can move to Alberta if they like those policies better. We are back to our discussion of exit and voice. We may want to push services to provinces if we feel we can get responsive government through both voice (since the government is smaller and closer to the people in the province) and exit (since Canadians can move to the province that most fits how they view life).

In addition, more decentralized decisions can lead to valuable experimentation. We do not know the best way to reduce emissions. As we saw in chapter 3, some argue that carbon prices are best while others argue that they are good but likely insufficient and that we need to think of other approaches. One of the oft-touted benefits of federalism is that it allows the provinces to engage in different experiments that can lead us to find better ways to do things. Such innovation is particularly important in the climate context, where there is so much uncertainty.[4]

Elinor Ostrom, a Nobel Prize–winning economist, argues that we cannot rely on any one level of government – global, national, provincial, or local – to deal with climate change on its own. She calls for a "polycentric approach" that involves action by all levels, in part because we need to experiment with different approaches but also because we cannot wait for an effective global agreement or even national action.

The risk from spreading responsibility across all governments is continued inaction. This takes us back to Hirschman's discussion of exit and voice. The problem with the provinces trying to work out a solution to climate change on their own is that while they may be able to agree on a cooperative plan where all do their part, such a plan would be vulnerable to exit by a province.[5] Again, it is the free rider problem – each province may see it as in its narrow self-interest to not act and allow other provinces to bear the costs of reducing emissions. If the federal government, on the other hand, tries to impose climate action on provinces, there is also concern about exit by provinces. Provinces may not agree to be bound by the federal government if they fear it will use climate change as a "Trojan horse" to take over more provincial powers.[6]

A possible solution may be a set of formal rules that allows for a rational division of responsibilities across the levels of government and limits the ability of one government to take advantage of another.[7] The Canadian Constitution is intended to set out the powers of both the federal and provincial governments. Does it provide the basis for rational action on climate change?

"Peace, Order and Green Government"[8]

The Constitution divides powers across the federal government and the provinces, and puts limits on what each can do. It gives to the federal government powers over broad issues such as banking, criminal law, fisheries, trade and commerce, navigation and shipping. The provinces look after issues that seem more local: hospitals (which has come to mean health care generally), municipalities, education, "property and civil rights within the province," and "generally all matters of a merely local or private nature in the province." They also have broad powers over non-renewable resources, forestry, and electricity. Both the federal and provincial governments can raise taxes, with the federal government's power being broader.

In addition, the federal government has a broad ranging power to make laws for the "Peace, Order and Good Government of Canada" (the POGG power), which is very important in the climate context. There are a few dimensions to this power. One is that the federal government has the ability to regulate matters that are of "national concern." As we will see, there are fairly stringent tests for a matter to be considered of national concern. Matters that have passed the test include aeronautics, radio, marine pollution and atomic energy. Another dimension of the POGG power is that the federal government can regulate in the case of emergencies like a war.

The courts have also developed rules for when there is a conflict. The Supreme Court has stated that cooperative federalism is a background principle – to the extent possible, courts are to read federal and provincial legislation so that they can both operate at the same time.[9] This principle cannot alter the distribution of powers but does point to how the powers are meant to work together and how federal and provincial laws should be interpreted. If both the federal and provincial government have the power to deal with an issue and their rules do not conflict, there is no problem.[10] When there is conflict, however, the key rule is "paramountcy." The federal rule prevails if there is a direct conflict, such as where the federal law says "no" to something and the provincial government says "yes." It also wins where the provincial rule does not directly say "yes" but in some way frustrates the federal "no" more indirectly.

Where does this framework leave the power to deal with climate change? The Constitution does not explicitly assign power over the "environment" to either the federal or provincial governments. Rather, the courts have decided that everyone gets power to deal with the environment: the federal government, the provinces, and even municipalities.[11] The Supreme Court has stated that the environment is a constitutionally difficult matter that "does not comfortably fit within the existing division of powers without considerable overlap and uncertainty."[12] Yet each government's actions have to fit within the existing set of powers. As two Supreme Court judges have stated, "Environmental protection must be achieved in accordance with the Constitution, not in spite of it."[13]

The provinces clearly have broad powers to take significant action on climate change. They can use their powers to regulate changes to buildings or to limit emissions from facilities or to put in place a cap-and-trade program. They can use their taxation powers to put in place a carbon tax. They can spend their money on subsidizing emissions reductions or energy-efficiency programs.

If the provinces do not act, can the federal government step in to either impose action or force the provinces to act on their own? The answer is in large part yes. The lack of clear boundaries around federal and provincial jurisdiction, though, leads to some ambiguity, and this ambiguity is the source of persistent problems. It is simplest to see the nature of the problem if we think about a few key climate tools: subsidies, regulations, and carbon prices.[14]

Beginning with subsidies: the federal government can give money to industry or individuals to reduce emissions or to provinces to put in place programs to limit emissions. It is a time-honoured way for the federal government to get things done in areas that may not be under its control. Universal health care is a key example. The provinces have power over health care, but the federal government provides money to the provinces if their public health care systems meet some conditions, such as universal coverage.

What about a regulation that limits the emissions of greenhouse gases or the content of fuel? Also likely yes. The federal government has the power to make criminal law. To be valid criminal law, the law must have a valid public purpose, like protecting health, and must have a particular form – it must prohibit something and the prohibition must be backed by a penalty. While this might not seem a natural source for climate change regulations, the Supreme Court of Canada has read the criminal law power in a way that gives the federal government a fairly broad power in the environmental area. For example, it held that the federal government could regulate substances like PCBs under its criminal law power.[15] The Court found that protecting the environment was a valid federal public purpose and, even though the federal government had set up a complicated regulatory structure, the regulation had the appropriate form for a criminal law. A straightforward regulation limiting emissions of greenhouse gases might then pass as criminal law on this same argument. In fact, the Federal Court has already upheld a federal government requirement that all diesel fuel produced, imported, or sold in Canada have at least 2% renewable fuel, finding it was aimed at dealing with the "real, measured evil" of climate change.[16]

How about the carbon price? A number of provinces challenged the federal government's ability to impose its backstop carbon price. This question had divided provincial appellate courts across the country, with two courts of appeal (Ontario and Saskatchewan) finding it constitutional and one (Alberta) unconstitutional. In 2021 the Supreme Court issued a 6–3 split decision on the constitutionality of the carbon price, with the majority finding it constitutional under the national concern branch of the POGG power.[17] Three dissenting judges would have found the federal carbon price unconstitutional. How the Supreme Court arrived at its conclusion has important implications for future federal climate action.

One question the courts considered was whether the carbon price is a tax. The federal government has a broad ability to raise "Money by any Mode or System of Taxation."[18] In the provincial courts of appeal, the provincial governments argued that the federal carbon price did not fall under this power because it was not aimed at raising revenue. Both the Ontario and Saskatchewan Courts of Appeal agreed that the federal government's carbon price did not fall under its tax powers.[19] The Supreme Court decision held that, from a constitutional perspective, the carbon price was not a tax, but rather constitutionally valid regulatory charges that advance the regulatory scheme of the act by altering behaviour.[20]

Another way the federal government could potentially obtain the power to address climate change is under POGG through its power to deal with emergencies. This power has been used in the obvious case of war, but also less obviously for extreme inflation in the 1970s. To fit under this power, the matter must be an "emergency," which seems to cover climate change, but also must be temporary. The majority of the Supreme Court did not consider the emergency branch of POGG as a basis for federal authority for the carbon price, possibly in part because of this requirement that the legislation be temporary. A decade or two is not long in the grand scheme of things but it may be in terms of the length of time the Court is willing to give sweeping powers over climate change to the federal government.[21]

The battle was mainly fought over whether or not the federal government's carbon price could be supported under the "national concern" branch of the POGG power. To be a matter of "national concern," it cannot just be that Canada as a whole will not be able to meet its international obligations, such as under the Paris Agreement. The Constitution does not give the federal government the power to implement international treaties – it can only implement treaties within its jurisdiction. To qualify as a "national concern," a matter must be of "sufficient concern to Canada as a whole"[22] and must be inherently national in character, meaning that "the matter is one that, by its nature, transcends the provinces."[23] By this measure, climate change would certainly appear to qualify.

However, the central question in this case was not whether *climate change* itself is a matter of national concern, but whether the specific subject matter addressed by the federal carbon pricing act is a matter of national concern. How broadly the courts should read the purpose of the act and defined its subject matter was a contentious issue dividing the majority and dissenting opinions in both the appellate courts and the Supreme Court. This framing was central to the decision because it in large part determined how significantly the legislation could be seen to impact provincial jurisdiction. The narrower that purpose – such as to establish a minimum price – the less likely it is to be seen as intruding on provincial powers, but at the same time the less it seems to be a matter of national concern. On the other hand, a broader purpose (addressing climate change or reducing emissions) is easier to see as being of national concern, but the likelihood of being found to impinge on provincial powers grows.

Ultimately, the majority of the Supreme Court found the matter of concern to be "establishing minimum national standards of GHG price stringency to reduce GHG emissions,"[24] which it found to be of sufficient concern to Canada as a whole to warrant consideration under the national concern doctrine.[25] In contrast, in his dissenting opinion, Justice Brown framed the matter of concern more broadly as "the reduction of GHG emissions."[26]

A key difficulty for the Supreme Court was allowing the federal regulation without undermining the basic division of powers – if the federal government has unlimited power to regulate in the area of climate change, the worry is it would essentially have an unrestricted ability to regulate everything: almost every act can be seen as having some connection to the climate. It could then take over broad swaths of provincial powers. This rests on a "transfer theory" of powers – that if the Court gives the power to the federal government under POGG, the entire power is taken away from the provinces.[27] Some, such as legal scholar Nathalie Chalifour, see this concern as overblown since such transfers either do not exist or rarely happen in practice, as in most cases both the federal and provincial governments continue to regulate.[28]

The Supreme Court has set two principal conditions on the federal government's regulation of matters of "national concern."[29] First, the matter must "have a singleness, distinctiveness and indivisibility that clearly distinguishes it from matters of provincial concern." A key test is the *"provincial inability test"* – what happens if the provinces fail to deal effectively with the issue? Second, the issue must have a *"scale of impact"* on provincial jurisdiction that is reconcilable with the fundamental distribution of legislative power under the Constitution.

There are two main approaches to the provincial inability test. The first approach, taken by the majority of the Supreme Court, points to the collective action problem that underlies climate change – that Canada needs a national response to climate change and the provinces cannot bring it about.[30] No single province can ensure that other provinces will not just keep emitting if it cuts back – as the Ontario Court of Appeal notes, if a province is harmed by climate change, "without a collective national response, all they can do is prepare for the worst."[31]

The majority of the Supreme Court emphasized the collective action challenge of establishing a uniform minimum carbon price across Canada without a federally established minimum national standard. In applying the provincial inability test, the majority found that the provinces cannot address, jointly or severally, the matter of establishing a minimum national standard because the failure of one or more provinces to cooperate, either by reducing its carbon price below the minimum standard or eliminating it completely, would prevent other provinces from successfully reducing Canada's GHG emissions.[32]

Furthermore, it held that such a lack of cooperation would have grave extra-provincial consequences. While the provinces *could* cooperate to establish a uniform carbon pricing scheme, the majority emphasized that the success of such a cooperative arrangement could be jeopardized by carbon leakage as businesses in high-emitting sectors move to provinces with less stringent regulation. Thus, the majority found the provinces unable to address the matter, noting that "in the absence of a federal law binding the provinces, there is nothing whatsoever to protect individual provinces or the country as a whole from the consequences of one province's decision."[33]

This first approach to the "provincial inability" test reflects the broader underlying problem – a collective action problem that hinders collective agreement, as opposed to a lack of capacity. The Supreme Court found support for such a broader basis for POGG in the international nature of the problem.[34] While the federal government does not have the power to implement international treaties beyond its already existing powers, the international nature of the problem and the existing international agreements supported the majority's finding regarding the need for the federal role.[35] The Court held that the predominantly extra-provincial and international character and implications of GHG emissions "support the conclusion that the matter at issue is qualitatively different from matters of provincial concern."[36]

The majority relied on trade and commerce jurisprudence from its decision on a federal plan to create a national securities regulator, noting that the national concern doctrine and the trade and commerce power pose "similar challenges to federalism."[37] Each of the provinces has its own laws and regulator for dealing with the sale of shares in companies. This duplication is seen as unnecessarily inefficient and costly, so the federal government tried to set up its own single national regulator. The Supreme Court, however, found that the federal government could not take over regulation of all aspects of the securities market but it could regulate "systemic risk" – risk that could spread across the entire financial system if not controlled. An individual province or even a group of provinces could not contain such risks.[38] While cautioning about the importance of not conflating the two powers, the Supreme Court held that the decisions in the securities cases offer "useful insight and are consistent with the modern approach to federalism."[39]

The analogy to climate change is not perfect, however. It is true that both systemic risk and climate change are public goods – that whatever one province does affects others and that all would be impacted by the effects.[40] However, systemic risk, like the pandemic, is more in the nature of what economist Scott Barrett calls a "weakest link" public good.[41] The risk depends on what all provinces do but the key is what is the weakest form of regulation, since that will determine if the risk to all is eliminated. Climate change, on the other hand, is an aggregate public good – it depends on the aggregate effort of all involved. In theory one or a small group of provinces could reduce emissions enough to allow Canada to meet its Paris targets. It is more costly, harder, and less fair than if all are involved, but it is possible.

The second approach to the provincial inability test sees the provinces as having the ability to address climate change if they want. This approach emphasizes the aggregate public good nature of climate change, arguing that provincial inability is established only when provinces do not have the jurisdiction to deal with a matter, rather than focusing on the collective action challenges associated with emissions reduction. The provinces can use their powers to address climate change and are taking some action. Moreover, they can cooperate if they see there are advantages.

On this view, it is not provincial "inability" but provincial "unwillingness" that is the crux of the problem.[42]

This is the approach adopted by Justices Brown and Rowe in their dissenting opinions for the Supreme Court.[43] Rowe J held that the extra-provincial consequences associated with a province's emissions, even when combined with evidence that provinces are not cooperating to reduce emissions, are insufficient to establish provincial inability. He argued that such an approach would mean that provinces are "unable" to legislate even in areas where they hold exclusive jurisdiction, because of the possibility of extra-provincial harm resulting from the exercise of provincial powers.[44] Instead, both Justices Rowe and Brown held that "the extra-provincial effects must be such that the matter, or part of the matter, is beyond the powers of the provinces to deal with on their own or in tandem,"[45] a condition that neither justice found to be satisfied.[46]

Thus, while the majority found the provincial governments to be unable to effectively *establish a uniform minimum carbon price across Canada*, Justices Brown and Rowe found the provinces able to *regulate GHG emissions*. This broader framing of the matter adopted by Justices Brown and Rowe supported their finding that the provinces *are* able to address climate change. Justices Brown and Rowe critiqued the majority's emphasis on the provinces' inability to legislate nationwide mandatory minimum standards as an indicator of provincial ability, arguing that this approach "renders 'provincial inability' meaningless by defining the matter in terms of 'minimum national standards,' something no province can do. By this logical sleight of hand, 'provincial inability' exists whenever Parliament provides for 'minimum national standards.'"[47]

The other main test that must be satisfied before the federal government can exercise powers under the national concern branch of POGG is the *"scale of impact"* test, which considers whether the impact on provincial jurisdiction is "reconcilable with the fundamental distribution of legislative power under the Constitution." Again, there have been two approaches. The first points out that climate change and greenhouse gas emissions touch on all areas of Canadians' lives and the economy. If the federal government can regulate in the area of climate change, the argument goes, it can regulate pretty much everything.[48]

Brown J, in dissent, adopts this view in arguing that finding the reduction of GHG emissions to be an issue of national concern would impact provincial jurisdiction on "a scale that is completely irreconcilable with the division of powers."[49] This understanding of the scale of impact on provincial jurisdiction is informed by the dissenting judges' broader framing of the matter of national concern. Brown J argued that if the federal government is found to have legislative paramountcy over such a broad area of concern, such a finding would extend federal jurisdiction to "the regulation of any activity that requires carbon-based fuel, including manufacturing, mining, agriculture, and transportation."[50]

The second approach to the scale-of-impact test focuses on the laws the federal government actually enacts. If the federal government can tailor its action to address the main issue with the collective action problem, such as through setting minimum pricing standards, then there will not be a large transfer of jurisdiction. The provinces can still do more if they want or do different things; the federal government is just setting the floor. Any problem with future attempts by the federal government to do more could be dealt with in future cases.

The majority adopted this approach, holding that, while provincial freedom to legislate less stringent carbon prices is restricted by the federal carbon price, this impact is limited and ultimately "outweighed by the impact on interests that would be affected if Parliament were unable to constitutionally address this matter at a national level."[51] The majority held that, given its narrow definition of the matter of national concern, the impact on provincial jurisdiction is limited to the regulation of GHG emissions pricing, and does not extend to regulation of GHG emissions more generally. Furthermore, because of the backstop nature of the act, provinces remain free to implement their own methods of pricing GHG emissions.[52] This qualified and limited nature of the impact on provincial jurisdiction gave comfort to the majority in finding the matter to be reconcilable with the fundamental distribution of legislative power under the Constitution.[53]

In constitutional terms, then, this is the challenge – we risk undermining either our ability to act or the powers of the provincial governments. Courts both for and against federal jurisdiction have been at pains to point out that the dispute is not over whether climate change is real or whether the federal policy is good or bad; rather, it comes down to a question of constitutional interpretation.[54] Legal scholar Jason MacLean sees it differently: these decisions are about politics more than constitutional law, and both sides are merely looking at how the federal government's plan fits within the existing political and economic structure. He and others like Chalifour argue, however, that federalism must be prepared to evolve, as we are in the midst of an existential fight.[55]

The majority of the Supreme Court tried to find some middle ground. In its decision on the federal government's attempt to establish a national securities regulator, the Court found that "a cooperative approach that permits a scheme that recognizes the essentially provincial nature of securities regulation while allowing Parliament to deal with genuinely national concerns remains available."[56] This is the balance that the Supreme Court tried to find in federal climate legislation – "cooperative federalism" as the way forward.[57] The majority emphasized that cooperative federalism requires courts to move beyond a rigid division of federal-provincial powers towards a "flexible view of federalism ... that accommodates and encourages intergovernmental cooperation."[58] While the majority was careful to note that a cooperative approach cannot override the balance of powers inherent in Canadian federalism, it found that the backstop nature of the federal carbon price achieved a cooperative balance between protecting provinces from irreversible harm due to the inaction of other provinces and leaving provinces free to legislate any GHG pricing system they choose as long as they meet the federal government's minimum standard.[59]

The Constitution, then, gives broad powers to the federal government to take significant action on climate change, but there are limits. As we have seen, some proposals for government action are sweeping, such as the Green New Deal in the United States, which covers not only climate change but income inequality, racism, and more. It would involve large federal government expenditures but also require federal regulation of a wide range of industries and activities. The United States, however, has a much more powerful federal government. In the 1930s, Prime Minister R.B. Bennett tried to introduce a similar measure to Roosevelt's New Deal policies, proposing reforming working hours, pensions, and the taxation system; giving money to farmers; and bringing in a minimum wage and unemployment and health insurance. However, the United Kingdom's Judicial Committee of the Privy Council, the final court of appeal for Canada at the time, held that most of these policies were outside of the federal government's powers.

So at least on a straight comparison to the New Deal legislation, the federal government in Canada seems less able to take strong action. On unemployment insurance, which the Privy Council found the federal government did not have the power to impose, Prime Minister Mackenzie King was able to get the agreement of all of the provinces to amend the Constitution to give the federal government explicit power over "unemployment insurance." It was a compromise position that required negotiation.

That will be part of our answer. The division of powers in the Constitution seems to be a key obstacle to climate action – we know there is a problem and we know there are solutions; we should just get it done and not have arcane debates over "powers." Again, though, on closer inspection, what seems like an obstacle may in fact help us find a path forward. Before we explore that potential, it is worth looking at how these rules have worked out in the climate context.

Going It Alone

The Supreme Court's "everyone's responsible" approach seems right in one sense. Canada needs to undergo a huge change and we cannot wait or rely on any one level of government. However, the diffusion of responsibility has not worked out that well for environmental law generally in Canada. Political scientist Kathryn Harrison called her history of Canadian environmental federalism *Passing the Buck* because of the tendency to shift responsibility to the other level of government when it is convenient to do so, with the result that nothing gets done.[60]

Unfortunately, neither has the diffusion of responsibility worked out well for climate policy. Very crudely, we can divide the history of Canadian climate policy into three broad phases. First, interestingly, there was a period of cooperation between the federal and provincial governments.[61] In the 1990s, and then later when the federal government was attempting to set the Canadian commitment under the Kyoto Protocol, it held lengthy discussions with the provinces. Unfortunately, the Chrétien government eventually took on a greater commitment under the Kyoto Protocol than the provinces had signed on to. This both eroded the spirit of cooperation and undermined trust between the levels of government.

The following phase was one principally based on independent federal and provincial action. After the Kyoto breakdown, there was no formal federal-provincial process to discuss national climate policy (with one partial exception in 1997), until very recently.[62] Most of the action came from provincial initiatives within their own provinces. The provinces tried to come to their own coordinated solution in 2008 and then prior to 2015 but failed to reach an agreement on a common approach or allocation of emissions reductions.[63] Subsets of provinces, however, did collaborate, such as Ontario and Quebec developing a common cap-and-trade system.

In the 2000s, the federal focus seemed more on developing its own plans, using spending to get provinces to buy in where necessary. Even the Conservative government under Stephen Harper, to the extent it acted, tended to use a unilateral approach rather than seek a joint plan with the provinces.[64] The federal government talked with the provinces about the specific individual measures, but to a large extent planning was limited between the federal government and the provinces and between the provinces themselves.

A third phase may have begun with the election of the Trudeau government in 2015, or at least this is how the Trudeau government would like to see it.[65] The Trudeau government and the provinces agreed to a framework for action in March 2016 – the Vancouver Declaration, which became the basis for the Pan-Canadian

Framework. Every province except Saskatchewan signed on to the framework, which set out a plan for carbon pricing as well as other regulatory action and subsidies. Prime Minister Trudeau has stated that his government was aiming to "work collaboratively" with the provinces to find solutions.[66]

Then the provincial governments changed. Conservative governments took power in Alberta, Ontario, and New Brunswick, and the appearance of togetherness fell apart. The federal government continued to claim it sought a cooperative approach. When it announced it was going to be imposing its backstop carbon price on reluctant provinces, the federal government highlighted its cooperative approach, noting that it had collaborated with the provinces over two years and had given them "flexibility to design their own climate plans."[67]

Political scientist Douglas MacDonald and legal scholar Jason MacLean argue that, while the federal government claimed to be taking a new cooperative approach, it was really just doing more of the same unilateral action.[68] The Trudeau government was able to reach an agreement because it was asking so little – it essentially required carbon pricing programs that the larger provinces already had put in place. And the federal government announced the decision to impose carbon pricing before negotiating with the provinces. The result has been a policy that is not ambitious enough to meet the national Paris targets and still led to federal-provincial conflict: the worst of both worlds. More recently, the federal government decided to significantly increase the carbon price, apparently again without much discussion with the provinces prior to its announcement.[69]

Wexit, Voice, and Loyalty

Why have we failed to find a solution when it is in all of our long-term interests to do so? Canadian constitutional rules have not been enough to create a system that provides the advantage of both the information and innovation of the provinces but also the efficiency, and perhaps national focus, of the federal government.

Both the federal and provincial governments have ample power to take strong climate action if they want to use it, but they have been reluctant.

We have discussed how Hirschman saw exit and voice as central to thinking about how organizations work. However, Hirschman also pointed to a third factor – loyalty. Loyalty can make parties less willing to resort to exit and also enhance voice. It allows members of an organization to take a longer-term view and be willing to bear some short-term costs. The idea is not that they are blindly loyal to the organization. Blind loyalty would mean you make no real contribution to improvement. Loyalty involves working within an organization to foster positive change.

One way to view the lack of coordinated action on climate change is that there is a lack of loyalty. To get all governments to work towards a common climate program, the federal government must not take advantage of the provinces and the provinces must not take advantage of each other.[70] They must all seek to further a national plan. Where does such loyalty come from?

One possibility, though it sounds strange, is that we require it. Some countries like Germany have a constitutional principle of federal loyalty enforced by the courts.[71] The federal government, for example, in exercising its powers to address climate change would be restrained from taking advantage of the provinces (such as by taking over too much power or reneging on promised funding) because it must act in accordance with this principle. This solution leans heavily on the courts to enforce limits on policy. Courts have to be willing and able to examine the contents of climate policy. If they are unwilling to do so, the principle will not support concerted action.

The other way to build loyalty is to cultivate trust. As we saw, Ostrom saw trust as a way out of collective action problems – cooperation arises where there is trust and reciprocity within a group.[72] In theory, all levels of government can effectively regulate in the area of climate change if there is trust that no individual government will free-ride or take advantage of the others. Trust would provide a foundation for each to understand and act on the joint responsibility, as opposed to providing the opportunity to shift responsibility when it is to their individual short-term advantage.

Unfortunately, trust and loyalty are in short supply in the climate context. Part of the problem may be historical, such as a lingering anger in Western provinces over the federal government's attempts to impose the National Energy Program (NEP) in the 1980s.[73] Trust is also eroding between the provinces themselves, as can be seen, for example, in the fights between Alberta and BC over the economic benefits and environmental harms of pipelines or the debates between western and eastern provinces over equalization payments and natural resources.

According to Ostrom, trust is built from some key elements. One is that the parties have to have information – on what the impacts of their actions are and on what everyone else is doing.[74] Lack of information seems not, fundamentally, to be the problem. The impacts of the rise in emissions are clear, and the federal government and each of the provinces in general have information about what each other is doing – about the policies each has in place and the level of GHG emissions in each province.

Ostrom argued parties must also be able to communicate with each other. Against this, MacDonald points out that the current process of federal-provincial negotiations is badly flawed:[75] it is too secretive and consensual and, combined with the fact that provinces can opt out at any time, the result has been that everything gets reduced to the weakest possible solution to guarantee agreement.[76] Part of the solution, then, might lie in a better process for negotiating federal-provincial agreements. Macdonald argues for an upfront agreement on a process of joint decision-making, annual meetings of first ministers, and ongoing coordination for environment ministers.[77] He views a clear negotiating process as key, citing the European Union as an example of a federation that has negotiated an arrangement in advance and so mitigated inevitable disputes.[78] Such a process could increase the commitment of all governments to meaningful action and aid in overcoming what Macdonald calls the West-East Divide. Jason MacLean also calls for a new form of federalism with a "new and collaborative partnership" between the federal government and other levels of government.[79]

In addition to processes and institutions for information sharing and communications, Ostrom contends trust also requires that members have the ability to sanction each other in some form and that each member care about their reputation for being a "trustworthy reciprocator." The federal government would need to experience some long-term reputational damage from taking on too much power or ignoring the provinces, while the provinces would have to feel some reputational hit from reneging on any deal they had with the federal government or other provinces. The potential impact of reputation comes in part back to the need for transparency – clear grounds for decisions make it harder to hide or to lie.

For Ostrom, groups can overcome commons problems successfully through trust. Such groups were able to find ways to trust each other to do things that seemed irrational in the short term but were optimal in the long run. Rules like the constitutional division of powers help, of course, but they are not enough. It is not enough for the Supreme Court to find that the federal government has power to address climate change. Trust allows governments to buy into common restrictions or, as discussed in the last chapter, to find ways to share the burden of climate action.

Returning to the Supreme Court's decision on a proposed national securities regulator, the Court stated that "we may appropriately note the growing practice of resolving the complex governance problems that arise in federations, not by the bare logic of either/or, but by seeking cooperative solutions that meet the needs of the country as a whole as well as its constituent parts ... Cooperation is the animating force. The federalism principle upon which Canada's constitutional framework rests demands nothing less."[80] The Court is saying that, even if we draw a clear line between the powers of one level of government and another, we need to do more to resolve tough problems. And climate change is one tough problem.

chapter ten

Fostering Trust

In the previous chapter we discussed the need to build trust within the Canadian federal structure. This chapter builds on this idea of trust, extending it to how governments make decisions. As we saw in chapter 7, we have not been able to deal with the central concern of Canadian environmental law – the heavy reliance on discretion and political accountability. While this can provide flexibility and decision-making based on expertise, it also allows space for short-term decision-making focused on economic growth and established industries.[1] As a result, Canadian targets and policies have been weak. Even when Canadian governments agree on some form of action – installing charging stations, setting up a cap-and-trade system – they often change their minds or do not follow up.[2]

Trust is the key to dealing with the problems of discretion – not the kind of trust that allows people to switch off and the public interest to lose out to private interests, but the other kind that engages people and gives them confidence that sound, reasonable decisions are being made. Trust will come from rebalancing the three main forms of accountability or, as we called them earlier, the three legs of the stool: political accountability, expertise, and legality. We need to (1) find ways to structure discretion so as to limit the bad accountability and increase the good; (2) enhance the role of expertise in decision-making; and (3) find appropriate ways to strengthen the role of the courts in limiting regulatory capture and short-term, poor decision-making. These are all interconnected in a self-reinforcing way if done properly.

Bolstering Good Accountability

We have discussed how there is "good" accountability and "bad" accountability. Good accountability connects decision-makers with those whose interests they are meant to consider – for our purposes, the public interest. Bad accountability ties decision-makers to those who are steering public decision-making inappropriately to their own ends. The difficulty lies in determining what "inappropriately" means in a given context. Currently Canadian environmental law leans heavily on political accountability but it lacks a solid foundation. We will not deal here with broad issues like electoral reform, but on more fine-grained areas where law and legal institutions can help to limit the bad and foster the good accountability.

Setting Targets. If broad, apparently open-ended discretion allows decision-makers to go in whatever direction they want and opens the door to bad accountability, one possibility is to restrict the paths forward. If we cut off particular policy directions, we may be able to stop at least the most damaging effects of bad accountability. Stronger, more specific environmental laws can limit discretionary space.[3]

A clear example is to try to specify particular outcomes in advance. A few provincial governments, for example, have stated emission reduction targets in their legislation. The BC *Climate Change Accountability Act* specifies targets of 40% below 2007 levels by 2030, 60% below by 2040, and 80% below by 2050.[4] It also requires the minister to set interim targets towards the 2030 goal. Alberta enacted the *Oil Sands Emissions Limit Act*, which sets a "cap" on oil sands emissions of 100 Mt per year.[5] There are exclusions to this cap, and there are no regulations yet in place to actually make it effective. However, in theory, it would constrain other choices made by the Alberta government.

The federal government has also recently decided to start setting firmer targets. Its general targets have been specified in guidelines – and so are non-binding. In June 2021, the *Canadian Net-Zero Emissions Accountability Act* came into effect, which sets a 2050 emissions target of net-zero emissions and requires the government to set national GHG emissions targets every five years starting in 2030.[6] It mimics the UK *Climate Change Act*, which requires the government to reach essentially zero emissions by 2050 and to establish five-year carbon budgets, although the UK is much more specific on how to set and alter the targets.[7]

Legislated targets provide some benefits.[8] They potentially give greater certainty to allow planning by companies and individuals. Targets also allow governments to pre-commit, which can reduce the probability that they will give in to later pressure for less action.[9] Moreover, targets can express society's aspirations, providing a signal to Canadians that reducing emissions is the right action to take in their own decisions on what to buy or what policies to support.[10]

Such an approach can, though, be less flexible and so fail to take sufficient account of new information, including on the need to make policies more stringent. The lack of flexibility may mean governments set weak targets in order to get agreement. Moreover, a narrow goal like an emissions target alone does not help articulate what trade-offs should be made – emissions can be reduced in a manner that makes society more equal or that increases inequality and places costs on vulnerable people. Targets themselves do not help with these choices. And of course, while rigid, targets can be changed or ignored – the Ontario and BC governments increased their targets and Alberta appears to have approved oil sands projects that would take it well beyond its 100 Mt cap.

Listing Factors to Consider. If the specificity needed is not a particular target (or at least not alone), an alternative is to set out factors that need to be considered in choosing the right policy. Legislation could specify that any climate policy decisions, for example, must be based on the best science or take into account social justice. There are many examples of the listing of such factors in environmental law. In order to limit discretion, for example, the federal *Impact Assessment Act* provides that Cabinet is to have the final say whether large projects are in the "public interest," but it must consider five factors: the contribution of the project to sustainability, whether the impacts of the project are significant, any mitigation measures put in place, the impacts on Indigenous peoples, and the effect on Canada's environmental obligations, including its climate commitments.[11]

In the climate change context more directly, the United Kingdom has adopted this approach, requiring that the government take account of a range of factors in setting five-year carbon budgets. These include the science and technology of climate change, economic circumstances including competitiveness of particular sectors, fiscal and social circumstances, energy policy, and regional differences.[12] Likewise, Quebec legislates not the target but a set of similar factors that the government should consider in setting targets.[13] The federal *Canadian Net-Zero Emissions Accountability Act* is more vague on what factors need to be considered, requiring that the minister in setting targets take into account the best scientific information available, Canada's international commitments with respect to climate change, Indigenous knowledge, and the advice of the advisory board established under the act.[14]

This strategy of specifying broader factors, however, may not significantly limit discretion. The factors are often vague and open ended. Using terms like "sustainability" without defining what they mean may do little to constrain environmental decisions. Moreover, there is generally no indication of how factors are to be balanced against each other. Cabinet in approving a project can point to different factors that still provide it with a broad space for decisions. Both the courts and the public have difficulty reviewing decisions because the language is so broad.[15]

On the other hand, factors may at least show the outer bounds, exposing decisions that should not be made if the relevant factors are considered. As well, requiring decision-makers to account for certain factors may actually change the decisions themselves. Forcing decision-makers to explicitly think about the effects on climate change, for example, could make them temper their choices. If, as we will discuss, decision-makers are required to directly address these factors in reasons, they may provide a more solid base for review by judges or the public. Again, while not a complete solution, this strategy may have some impact on ultimate decisions through strengthening both the decision-making process and the ability to review it.

Setting Broader Goals. One step further would be to set broader goals about how all decisions are to be made. Legislation could specify, for example, that all government decisions have to be made with social justice in mind. While vague, this ties in with a discussion we had earlier about a "just transition." Recall that Sen argued for basing policy on what it takes to ensure that people can live lives they have reason to value. He states: "Without ignoring the importance of economic growth, we must look well beyond it."[16] Another Nobel prize–winning economist, Joseph Stiglitz, has emphasized how GDP alone is a poor metric. A focus on GDP growth misses out on things we care about, like resource depletion. The policy lens needs to be broader. For Sen, it would include education, health, political freedoms, protection against unemployment, and more.

The difficulty is, of course, that these criteria for discretion become even more vague and so even less of a guide. Some attempts have been made to move these ideas into a more structured form, such as measures of well-being.[17] New Zealand has brought these measures at least partially into their decision-making process by putting in place a "wellbeing budget"[18] with new spending focused on mental health, child poverty, Indigenous peoples, transitioning to a low-carbon economy, and digital issues. Perhaps as importantly, the government seeks to use a wide variety of measurable indicators to gauge their progress. Boyd has suggested similar broad goals to aid in improving Canadian environmental policy.[19]

Nova Scotia seems to be taking steps in this direction with its proposed *Sustainable Development Goals Act*, which sets a goal of "sustainable prosperity" that it defines as including *Netukulimk* (an Indigenous concept related to responsible use of resources), sustainable development, and a circular and inclusive economy.[20] If combined with metrics to make at least some aspects measurable, such an approach could help support accountability of elected officials to the public.

Involving the Public. Even without specific rules or criteria to balance, good accountability can be improved through enhancing the role of the public in decision-making. The public may, for example, become more involved or concerned if decisions are more transparent. Increasingly, environmental legislation (and courts more generally) requires decision-makers to give reasons for their decisions. Such transparency may have positive impacts on decisions where the government has to justify itself to the public. The federal *Canadian Net-Zero Emissions Accountability Act* attempts to leverage transparency by requiring the government to issue periodic progress reports to Parliament and the public.

One step further is to get the public involved in the decision-making process itself. In part, the public is starting to become involved on its own in trying to influence climate policy through large demonstrations. While such demonstrations signal strength of conviction of part of the population and can put pressure on politicians, they also reveal a failure in current forms of participation and accountability: reaping the benefits of public participation requires more fine-grained solutions to involve the public when decisions are actually being made.

Great strides have been taken over the past forty years in involving the public in environmental decisions, including ensuring notice of proposed decisions, allowing comments in many situations, and holding public hearings for larger decisions.[21] Such participation can provide information to decision-makers about the benefits and costs of decisions and can allow the public to express their views on the trade-offs involved. These processes also provide information to hold public officials to account. Moreover, as we will talk about next chapter, if individuals' values are not fixed, public participation processes may help shape common viewpoints through deliberation.

Transparency and participation are, again, necessary but not sufficient. We know that the public's attention generally waxes and wanes, and this is particularly true across individual decisions like mines or pipelines. The public may not care at all about very important decisions (or know enough to care) and may focus too much attention on other, less important ones. Winfield, for example, has argued that the reason many of our individual environmental assessment decisions have been so contentious is that the public has focused their anger and concern on a few representative projects.[22] Decisions on these projects then get swamped by public

participation that would be better focused on getting overall policy right. Further, as we pile on participation requirements, the
process of change can become slower and costlier. As we discuss
next chapter, in some cases it may even lead to worse decisions.
Individuals can make mistakes in thinking about some of these
risks because of a variety of issues with how we process risks, such
as where there is a small probability of large harm.

Legal scholar Jason MacLean sees public participation as central
to restoring public trust – he calls for "bidirectional and responsive
engagement with all Canadians, not just special interests," along
with greater collaboration with provinces and Indigenous peoples
and more evidence-based policymaking.[23] We need to enhance the
good aspects of public participation while limiting the downsides.
The balance is crucial but hard to find.

Enhancing Expertise

In addition to strengthening good accountability, increasing expertise in climate policy-making can lead to better decisions. We
turned to experts once COVID started to hit. Few people were asking whether politicians should be listening to health profession-
als on the issue of social distancing or how to "flatten the curve,"
although political considerations increased in importance as time
went by. Expertise is similarly central to climate change policy.[24]

Government could set up an independent body to advise it
on how to make decisions on climate change.[25] Such arms-length
bodies may garner more trust from the public and may be better able to resist industry pressure by identifying effective policies.
Involving expert bodies in decision-making has a relatively long
history in Canadian environmental law, such as in environmental
assessments.

The value of independent bodies can depend on how they are
incorporated into decision-making. The federal *Species at Risk Act*
has a "reverse onus" requirement for deciding which species should
be listed as at risk.[26] The act's protections come into play only once
a species is listed. When the act was being created, environmental

groups wanted a species listed if an expert scientific body (called COSEWIC) recommends its listing. Others wanted Cabinet to have complete discretion on listing, only taking advice from COSEWIC. The drafters of the act came up with a creative middle ground. Under the act, if COSEWIC recommends listing a species Cabinet then has discretion whether to list, but if they decide to not list the species they must issue reasons why they are not following the scientific recommendation. Moreover, if Cabinet does nothing within nine months the species gets listed.

The result is an interesting attempt to combine scientific expertise and political accountability. Requiring Cabinet to give reasons for not listing does not guarantee an outcome but connects to public accountability. Most species recommended by COSEWIC are listed, although some species have not been listed, often due to the economic impacts of listing. Unfortunately, greater discretion seeps in when it comes to actually protecting species through recovery plans or issuing orders to safeguard species in real trouble, and the record there has not been as stellar.

The United Kingdom has attempted to draw in scientific expertise in its decision-making on climate change. Its *Climate Change Act 2008* creates a Committee on Climate Change to provide analysis and advice on how to meet the UK's carbon budgets, including on the allocation of the budget across sectors. The committee, whose members are appointed by the government, must be composed of members such that it has "experience in or knowledge of," among other things, business competitiveness, climate policy and science, economic analysis, and energy production.[27] The government has to consider the committee's advice when setting the carbon budgets; the committee has to give reasons for its advice and publish that advice. Moreover, it must report to Parliament each year on how the UK is doing regarding its budgets, and the government has to respond to this report. The Canadian Institute for Climate Choices notes that the committee is seen as "a credible, authoritative, and trusted voice on climate change – and its advice has been broadly accepted by governments," but the willingness of governments to follow its advice may become weakened as the decisions impose greater constraints on the economy when budgets get tighter.[28]

A central concern with such a solution is ensuring that the "independent" body is truly independent. Canadian governments set up tribunals and boards and call them "independent," but in fact it is often Cabinet that hires and fires the people on them. Ron Ellis, a labour law scholar, conveys his opinion of government tribunals in the title of his book: *Unjust by Design*.[29] Credible independent advice results when the body is seen as actually independent. The National Energy Board, for example, lost credibility with some who felt its members were biased and overly industry focused.

Governments exercise control over the degree of independence in a variety of ways. They can set conditions on who can be on the body (such as having particular types of expertise and a lack of conflicts of interest) and how they decide who can sit on the body. They can also grant some job security, and hence some independence, through the length of time a person sits on the board and the conditions for removing them. The governor of the Bank of Canada, for example, is a position that requires considerable independence from the government in order to give markets confidence in his or her decisions. The *Bank of Canada Act* requires that the governor have financial experience and no conflicts of interest, be appointed for seven-year terms (so more than the life of most governments), and have protections on removal from office.

A significant weakness with relying on experts, even if independent, is the legitimacy of value choices they make. As legal scholar Jocelyn Stacey notes, we "assume that objective expertise can provide the constraints on discretion that the legislature is unable to provide" and "an independent decision-maker is simply doing what the legislature, or indeed anyone, would do if they possessed the requisite knowledge."[30] It assumes an objectively right answer, which is often not true given the uncertainty underlying climate change.

Canadian governments have tried different ways to bring independent or expert advice into their climate change planning. Before the approval of large projects, the federal *Impact Assessment*

Act requires an expert panel report to Cabinet on the effects of the project. To meet its targets, the BC *Climate Change Accountability Act* specifies that the government is to establish an advisory committee of up to twenty people, half of whom have to be women, which must include at least one representative of Indigenous peoples, local government, environmental organizations, academics, unions, persons living in rural and remote communities, and the business community.[31]

The federal government's *Canadian Net-zero Emissions Accountability Act* establishes an "advisory body" appointed by Cabinet to provide advice on achieving the targets. While the initial draft of the bill did not require any particular expertise for this body, the final act requires the minister to consider "the need for the advisory body as a whole to have expertise in, or knowledge of" a range of issues, including climate science, Indigenous science, relevant social sciences, energy, and relevant technologies.[32] The act relies on the requirement for reason-giving to hold the government to account if it fails to follow the advice of this body. In addition, the government's progress is to be reviewed by the federal Commissioner of the Environment and Sustainable Development, an independent body, every five years.[33]

Canadian governments, then, have started to enhance the role of independent experts in climate policy. This development should be extended more broadly throughout the process, with attention to ensuring these experts are in fact independent.

Strengthening the Role of the Courts

The third leg of the stool is the courts – can they help support better climate decisions? We saw in chapter 7 how there were essentially three main intersections between the courts and climate policy (aside from the federalism issues we talked about in the previous chapter): constitutional rights generally, the constitutional duty to consult and accommodate Indigenous peoples, and administrative law (non-constitutional challenges to government decisions on the basis that they either were made following unfair procedures or are in some sense unreasonable or incorrect). Each has a potential role to play.

Constitutional rights generally have arguably the least role at this juncture. We saw in chapter 7 how existing Charter rights have not so far been a significant constraint on climate policy, though they have potential to raise public concern. An alternative would be to create a new constitutional right – but a right to what? A healthy or clean environment? A stable climate? Intergenerational equity? The difficulty is in defining a right that is specific enough to allow courts to usefully review government decisions but not so specific ("clean water") as to be impossible to fulfil. Legal scholars David Boyd and Lynda Collins argue that Canada is behind other countries in having no such right; a majority of UN members recognize some form of it. An environmental right could arguably provide a safety net to hold governments to account for their decisions, promote participation by Canadians in climate decisions, and foster pro-environmental norms among Canadians by signalling the importance of the environment.[34] It may be a way of providing stability or smoothing out the waves of environmental law by stopping the backsliding when the economy declines and public attention wanes.[35]

Beyond the trouble in defining the right and the general difficulty of bringing such litigation (it is slow and costly), it would be very hard to get such an environmental right enshrined in the Constitution. That document, not surprisingly, is difficult to amend and, as befits a constitution, there have been few changes over time. There would be significant resistance from those who would lose by the change, including some industries. Many scholars and environmental groups have been pushing for an environmental right of some form. Others such as Jason MacLean argue that the political backing is not yet sufficient for it to aid with climate change.[36] Regardless of what you feel about this project, it is at least a few years down the line.

Legal scholars Lorne Sossin and Lynda Collins have suggested a related way to control discretion – have courts invoke a new *unwritten* constitutional principle of "ecological sustainability."[37] The Supreme Court of Canada has found a few unwritten constitutional principles that underlie the written Constitution,[38] among them federalism, separation of powers, parliamentary sovereignty, and the rule of law. Sossin and Collins argue that an unwritten constitutional principle of ecological sustainability could help courts to constrain the discretion exercised under environmental legislation and to interpret the legislation itself. It could be a "more consistent and compelling framework for guiding environmental discretion, a mechanism for enhancing reconciliation with Indigenous peoples, and a catalyst for political, social and cultural change."[39]

An unwritten constitutional principle is certainly more attractive than a new right in terms of its immediacy. Courts could recognize it immediately and apply it to climate decisions. One difficulty, of course, is that judges would have to be bold in taking on a broader role in reviewing decisions based on a principle they found underlies the Constitution. Unwritten constitutional principles have been controversial, as they raise the issue of the legitimacy of judicial power to create new categories of constraints on the legislature or government decision-makers. Judges have not tended to be overly adventurous in this area, and the Supreme Court recently has shown some reluctance to lean on unwritten principles.[40]

Some provincial governments have implemented "rights" in the environmental area, but these tend to be procedural – giving the public some rights to notice of decisions impacting the environment and to comment on these decisions.[41] These environmental bills of rights, or other substantive environmental statutes such as requiring environmental assessment, may also include a requirement for decision-makers to provide reasons. These procedural rights might not have the bite of a well-defined, respected substantive environmental right (like a right to a healthy environment), but they nonetheless provide an important form of accountability and participation. The federal government has recently proposed adding some (unclear) form of a right to a healthy environment into the *Canadian Environmental Protection Act*, though the language around the right is heavily qualified by consideration of social, economic, health, and scientific factors.[42]

The second role for the courts – that of enforcing the Crown's duty to consult and accommodate Indigenous peoples prior to making decisions that could affect their rights or claims – is more significant. Teasing out the connections between the duty and climate change action and discussing reforms would require another book. For our purposes, a central area of overlap lies in the impact of the duty on how government decision-makers exercise their discretion. Decisions about resource projects such as new mines or pipelines, for example, cannot be legitimately made without fulfilling this duty. It is important not only for how we think about projects that may increase emissions but also, going forward, for those that are aimed at helping reduce emissions. The federal government views the path to net-zero as lying through increased exploitation of Canada's minerals, such as those necessary to produce batteries for electric vehicles. Decisions on new or expanded mines will need to conform to the constitutional duty to consult and accommodate affected Indigenous peoples.

The federal government's implementation of the UN Declaration on the Rights of Indigenous Peoples (UNDRIP), which refers to the "free, prior and informed consent" of Indigenous peoples, raises issues about the meaning of "consent."[43] But even beyond the controversy over the meaning of consent, the courts will need to re-consider underlying assumptions on which the duty is built. Courts defer to "reasonable" government processes since, as is understandable, the duty arises in complex situations and involves a wide variety of considerations about which judges are not expert. However, this deference needs to be more finely attuned to the underlying biases in the legislation and structures to which the judges are deferring.[44] In at least some circumstances, moreover, the Supreme Court appears to lack a nuanced view of the relationship between the Crown and tribunals assessing the consultation and accommodation. Where a tribunal making a decision that is subject to the duty is not actually independent of the government, courts should be less ready to defer to that tribunal's decision that the duty has been met.[45]

Finally, the courts also have an extremely important and increasingly controversial role to play in reviewing government decisions even if the Constitution is not invoked. As we saw in chapter 7, the Supreme Court emphasizes the need for judges to defer to government decisions such as whether to approve a pipeline or to regulate emissions from cars. Such deference creates space for government actors to use their discretionary powers in inefficient, ineffective, or unfair ways.

Two main approaches have been suggested to enhance the role of courts in reviewing environmental policy. First, judges could take a closer look at the decisions. Legal scholars Chris Tollefson and Jason MacLean, for example, argue that a stronger role for the courts is "a critical precondition of timely and efficient" environmental assessment processes and of progress on climate change.[46] Such a role would help restrain arbitrary or biased government decisions.

The risk from this approach is that judges will be doing more harm than good. It is not clear judges would be good at reviewing decisions in areas of great uncertainty such as climate change – and in fact some have argued that in such cases, judges have to keep their hands off these decisions as the risk of their getting things wrong is so high.[47] Moreover, judges may be tempted to decide in accordance with their own preferred policy, which raises questions of their legitimacy to make these decisions. Tollefson and MacLean argue that the courts are able to deal with the technical and scientific basis for environmental decisions, as it is not that different from what they do in the other areas such as patents, procurement, and competition law. They contend that "robust judicial review will not only help restore Canadians' trust in the government's environmental assessments and regulatory processes" but also reduce litigation.[48]

The alternate approach focuses less on judges taking particular stands on government decisions and more on requiring decision-makers to justify their decisions. Jocelyn Stacey, for example, argues that we cannot rely on a formal view of the role of the courts nor on specific rules or independent expert decision-makers. Instead courts should require decision-makers to provide "adequate justification" for their decisions.[49] Decision-makers must provide enough information and reasoning to give the public the basis to "hold the elected officials ... to account."[50] Others, however, argue that reasons do not offer enough of a constraint and still allow potentially unlimited power to government decision-makers.[51]

In part, the Supreme Court draws a bit from both approaches in its *Vavilov* decision, its most recent broad statement of the role of the courts in this area. It focuses on the need for the courts to pay close attention to the reasons given by the decision-maker and to undertake "robust" review of government decisions.[52] The difficulty is in figuring out how courts can play this role. It is not enough to just require "reasons" from decision-makers, since to make this a meaningful requirement, courts would have to see if the reasoning was in some sense adequate. The Supreme Court has struggled over the years with finding a way to have judges review reasons of decision-makers without substituting their own views of the right answers. In *Vavilov*, it tries again to give clear guidance to judges on how to walk this line, with advocates of both approaches fearing it has not got the balance right.

Interestingly, the Supreme Court addressed the issue of discretion in its decision on the constitutionality of the federal *Greenhouse Gas Pollution Pricing Act* (GGPPA). The dissent pointed to the extreme breadth of discretion under the act as an indication that the federal government was not actually setting a standard for pricing. For example, Cabinet was given the power to impose the federal pricing scheme on any province based on the provincial plan's "stringency," which was not defined in the act. The discretion in the act was so broad, and backed by a clause that allowed for regulations made by Cabinet to prevail even if they conflict with the act, that the dissent argued it did not constrain Cabinet in any meaningful way and could lead to Cabinet, for example, imposing pricing that shut down the oil and gas industry. The majority, however, was not concerned. It stated that while there was a broad grant of discretion, any decision of Cabinet "would be open to judicial review to ensure it is consistent with the purpose of the GGPPA and with the specific constraints" under the act.[53] The Federal Court subsequently decided a challenge by Manitoba to the federal Cabinet's regulation applying the federal pricing scheme to it. The Federal Court, noting the high bar for courts' overturning regulations, upheld the regulation.[54]

In part, the debate about the role of the courts is political. In the United States, for instance, the courts' main approach is deferential to government decision-makers; this deference was pushed by conservatives in the 1980s because they worried the courts were too liberal and promoted left-leaning policies. Recently, liberal commentators have defended a deferential approach from attacks by conservatives who want judges to take a stronger role in reviewing government decisions.[55] Expectations of whether judges tend to be aligned with particular policy views may then influence feelings about the appropriate role of judges. For those who want strong climate action, a push for judicial deference requires a belief that government decision-makers will make, on balance, good decisions; that judges will make more mistakes than they correct or will be biased against action in their review; and/or that the other legs of the stool (expertise and political accountability) are sufficiently strong that a minimal role for judges can support strong action.

Such legal constraints also require judges to be willing to take on this role. Jason MacLean points out that the federal government tried something similar to the *Canadian Net-zero Accountability Act* before. The *Kyoto Protocol Implementation Act*, introduced as a private member's bill at the time of Prime Minister Stephen Harper's minority government, required the federal government to put in place plans to ensure that the government met its Kyoto commitment. It specified considerations to be included in the plan as well as required reports to Parliament and the public on progress on the plan and review by an independent body. Environmental groups took Harper's government to court when it proposed plans that would clearly not meet the Kyoto targets. The Federal Court held that the matter was non-justiciable – that is, was not a matter for the courts but in fact depended on other forms of political accountability.[56] Courts may well take a similar position on the proposed federal accountability structure.

Levelling Off

We have a range of choices for improving Canadian institutions to bring trust back into climate policy. Our major elements – political accountability, expertise, and legality – are "imperfect alterna-tives."[57] Each has its strengths and weaknesses. Fortunately, the weaknesses of one can be offset by the strengths of the others. Adopting clearer targets or rules for decision-makers not only makes it easier for the public to hold elected officials to account but improves both the ability and legitimacy of stronger review by the courts. Increasing the independence of experts improves policy by supporting better decisions by political actors or judges. More-over, there are feedback loops. A stronger role for the courts can induce elected officials to be clearer in their legislation to ensure their wishes are followed. A demand by the courts for reasons can improve the reliance of political actors on expertise or improve their decision-making processes.

Such changes, though, are necessary but not sufficient. Just as we will not find the solution to climate change solely in the tech-nical question about what tool to use, institutional design is only part of the answer. Regulatory capture, or even just more basic political calculus, limits the ability to make these or other changes. If discretion is swayed by private interests, why would we believe that targets, either specific or more general, will be set in a way that actually constrains that discretion? It will take stronger support for change. For that we need those who believe in the goals of climate policy, the Baptists in addition to the Bootleggers, and to that we finally turn.

chapter eleven

Setting the Foundation

Bringing about the changes we discussed in the last few chapters as well as opening the door to the use of the tools we discussed in chapters 3 and 4 requires a solid base of support from individuals who think significant climate action is necessary. As we will see, individuals can make a difference by altering their choices, such as about the cars they drive or how they heat their homes. The bigger change, though, likely comes from their political choices. Economist Mark Jaccard, for example, argues that a key move is to get people to vote for politicians who are "sincere" about climate action.[1]

There has been an increase in support for stronger climate policies in recent years. However, it is not enough. We have seen how environmental issues have been swamped by short-term economic concerns in Canada in the past, and the pandemic has brought with it plenty of those concerns. Moreover, the changes are so large and needed so urgently that the worry is that the support is still not sufficiently deep. Sharp spikes in energy prices in Europe, for example, raise fears of a backlash against climate action.[2] We need stronger norms around climate action.

This chapter examines what role law and institutions can play in developing these norms. It in part seems like a chicken-and-egg problem – we need strong climate policies to get support for strong climate policies. However, there is more to it. We need to think about what policies are put in place and how they are chosen. To understand the connection, we first need to discuss what we know about peoples' choices and norms around climate change.

Individuals and Climate Change

We have focused in much of the book on what government can do to reduce large sources of emissions – reducing emissions from oil and gas production or electricity generation, for example. Individual choices about driving, home heating, and even diet are also a significant source of emissions.

Taking emissions attributable to households broadly, households account for 42% of Canadian GHG emissions, with about 25% of that due to transportation, 20% due to heating, lighting, and appliance use, and about 15% due to food and beverages.[3] We can break these emissions down into two groups. There are the "direct" emissions caused by individuals, such as the emissions from driving their cars or heating their homes. Some include in this category the electricity used by a household. There are also "indirect" emissions arising from industry producing goods and services individuals purchase for their own use.[4]

Direct Emissions

Most direct emissions come from transportation and heating homes. In 2015 direct emissions (including both transportation and home heating) by Canadian households accounted for 19% of all emissions in Canada.[5] Overall, transportation was one of Canada's largest sources of emissions in 2017 at 24% of total emissions, rising between 1990 and 2010 but becoming stable since.[6] The Pembina Institute estimates personal driving amounts to about 10% of Canada's emissions.[7] Emissions depend on what you drive[8] and where you live. A Pembina report on driver types and differences in emissions found a suburbanite emits about twice as much as a metro motorist, but the average rural driver has fewer emissions than the suburbanite, in part because it was assumed they do not commute as far to work.[9]

In Canada over half of residential emissions in 2018 were from space and water heating.[10] Residential emissions (such as home heating) have declined since 2005, with efficiency offsetting increased population and floor space.[11] Electricity use by households made up about 4% of Canadian emissions.[12] As we have seen, these emissions vary greatly by province. Alberta, Saskatchewan, and Nova Scotia all use some coal in their energy mix, which leads to higher emissions per amount of electricity generated. Quebec, by contrast, is mostly hydro, and Ontario has fairly "clean" electricity generation from a fossil fuel perspective, as it relies heavily on hydroelectric power and nuclear.[13]

Indirect Emissions

'Indirect" emissions are produced in making the things we buy and use. Some attempts have been made to estimate "consumption-based" emissions – looking at all the emissions associated with activities undertaken or goods and services purchased by households (regardless of where the actual emissions occurred).[14] Economists Sarah Dobson and Kent Fellows estimated the consumption-based emissions across the country and compared them to the usual approach of estimating the emissions actually released in a province.[15] They use the example of a good produced in Ontario using electricity generated from natural gas from Alberta that is ultimately sold to a consumer in Quebec. Under the usual approach, the emissions from the production of the natural gas would be allocated to Alberta and from the electricity generation and production to Ontario. No emissions would be allocated to Quebec if the use of the product did not result in emissions. The consumption-based approach attributes all these emissions to Quebec.

The difference between these approaches is stark. Figure 11.1 compares consumption-based emissions on a per capita basis for households, firms, and government to production-based emissions (the usual approach). As we have seen, if we focus just on where the emissions are produced, Alberta and Saskatchewan stand out due largely to the oil and gas industry. However, when we look at the emissions in what Canadians consume, Alberta, Saskatchewan, and Nova Scotia remain at the high end while Ontario, Quebec, and Manitoba are lower (in part due to electricity generation again), but the emissions become much more similar across

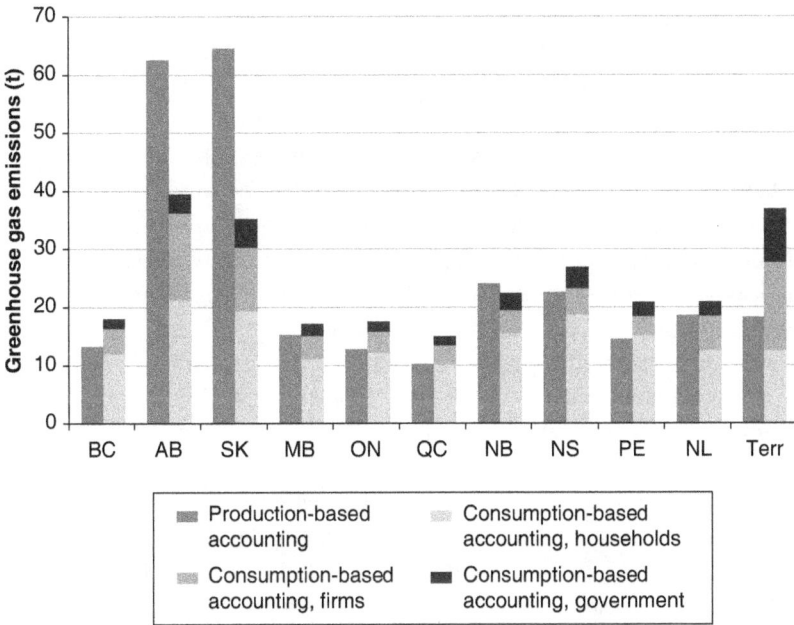

Figure 11.1. Per capita provincial emissions by production and consumption-based approach.

Source: Dobson and Fellows, "Big and Little Feet: A Comparison of Provincial Level Consumption and Production-Based Emissions Footprints," University of Calgary School of Public Policy Publications 10, no. 23 (2017), fig. 4.

the country. Some emissions are attributed to provinces such as Ontario or Quebec when they consume oil from Alberta's oil sands.[16] Note that this figure does not show the considerable emissions that arise from goods imported from other countries, such as food grown in countries like Mexico or products from China.[17]

Market and Political Choices

Individuals' choices about what to buy or how to live, then, have a significant effect on Canada's GHG emissions. Individuals can make the greatest difference by reducing emissions in a few key areas: transportation; home heating, lighting, and appliances; and food.[18] In the US context, three of the top seven adjustments relate to driving: purchasing a more fuel-efficient car, carpooling, and adjusting driving behaviour.[19] In terms of your home, electrification can make a big difference depending on where you live. As burning fossil fuels such as natural gas for heating emits greenhouse gases (and can worsen air quality within the house), moving to electric heat can have a big effect in provinces where generation is low emitting (such as in Quebec and Ontario). Moreover, individual choices have valuable knock-on effects. They provide the basis for private (and public) investment that can drive innovation.

However, as with much of climate change, there are dueling perspectives on how much we can rely on individuals changing their choices. On the one hand, some argue we as individuals have to make fundamental, life-altering choices if progress is to be made on climate change. Governments are part of the answer, but really nothing is going to happen until we act. This view has led to whole industries around these alternative choices.

Others as different as Naomi Klein and Mark Jaccard argue that while we should probably do all these things, they will be too little, too late. We need to take more societal action. Naomi Klein in her recent book *On Fire* writes: "The hard truth is that the answer to the question 'What can I, as an individual, do to stop climate change?' is: nothing."[20] Jaccard agrees that it really is not individual action as consumers that is going to make the big difference; government regulation and technological change are necessary in a few key areas.[21] Such government action requires political support.

Do people appear to care enough about climate change to change their choices about what they buy or how they act or how they vote? Polls indicate that climate change is an important issue for Canadians. It is up there with the economy. Polling during the 2019 federal election showed about 24% of Canadians saw the environment and climate change as the top concern while only 14% chose the economy.[22] Canadians still appear concerned about climate change even during the pandemic and the resulting damage to the economy.[23] Most Canadians in the fall of 2021 expressed a willingness to pay more (such as through taxes) to reduce Canada's emissions, although it was only a "slim majority."[24] Moreover, in the most recent federal election even the Conservative party, which had long been opposed to carbon pricing, included it as part of their climate plan.

However, the environment has been at the top of people's priorities list before (as recently as just before the financial crisis), only to fall when times get tough. There are warning signs. In that poll in fall 2021, about 40% said they were unwilling or somewhat unwilling to pay more to reduce emissions.[25] Even when concern seems high, things are a little less clear when it comes to changing individual behaviour. Yet this support is not independent of environmental policies. To see the connection, we first need to think about how people choose across their options.

How People Choose

People make choices about such things as the car they drive, the food they eat, or the way they heat their home for three broad reasons. First is price. We can think of price narrowly as the price of a good such as a car – electric cars are more expensive so on a purely financial basis fewer people are willing and able to buy them. But we can also think of the price of a choice more broadly as all the costs associated with it, including not only the price of the car but also the longer-term costs to ourselves or our children or people in other countries from the continued use of fossil fuels. If we think the problem is that people do not value these costs fully, this can point to one set of policies based on changing these costs or how people view them. Governments can subsidize electric cars, for example.

However, we are not driven just by price in all our decisions. A second basis for our choices is how others will view those choices. We value our reputation with our friends, co-workers, and neighbours. I may worry that if I make certain choices, others will think less of me or will be less likely to hire me.[26] Alternatively, making certain choices may make people think more of me. This concern about what others think can have a big impact on our decisions and can create patterns of behaviour and belief that are called "social norms." These social norms can steer our choices in many cases, but they do not work for everything. For social norms to affect my choices, there needs to be some agreement on what is the right choice as well as some ability for others to both be able to tell which choice I made and to impose some penalty on me.

These forms of social norms may work for actions like littering. If your community agrees that littering is bad, you may be less likely to at least overtly throw your garbage on the ground, as others will see you do it and may speak out. Even then there are limits – you may throw garbage down if there is no one around or if those around you are not people you know or care about. Social norms may also work for some climate-related decisions. My neighbours may know if I drive or bike to work. However, there may be enough uncertainty around some choices that reputation will not work well. There could be some confusion, for example, on the best type of car to drive – does it take more emissions to produce a plug-in electric car as opposed to a regular car? To the extent there is uncertainty about the right choice, norms are less likely to be effective. Moreover, there are many important individual decisions that others cannot even see you make – how do others know how you heat your home or whether you eat red meat on a regular basis? And of course social norms can point to non-climate-friendly choices just as well, such as a norm in some areas or groups to drive large SUVs.

The third reason that people do things is because they feel they are the right thing to do, even if there is no personal benefit. This sense of right and wrong may mean certain choices are not even an option – you would not steal, you would not kill someone. For an economist, if you do these things, you still pay a price – guilt. We can think of these norms as "internally" enforced since you punish yourself, unlike the other norms that are "externally" enforced by others' opinion of you. These "internally" enforced norms may be tied to how people see themselves. If a person views herself as an "environmentalist" then this identity may be tied to certain choices – not to drive or eat red meat. She may feel guilty if these choices are not followed.[27]

People, then, may choose to take climate action because of the price, because they care about what others think, or because they feel it is the right thing to do.[28] There is a potential connection between these motivations. Some studies have shown that putting a price on something – a fine for littering, a payment for giving blood – may not actually lead to more of what you want (less littering or more blood donated). The monetary price may "crowd out" peoples' social cooperative behaviour.[29] In a famous example, when a daycare put a price on picking up the children late, the number of parents who were late actually went up. The parents no longer felt bad about being late; they just paid the price.[30] Such crowding out does not always occur, nor is it always large.[31] However, the potential for crowding out means we need to carefully consider how whatever climate policies we put in place will affect other reasons for action, and look for areas where the tools we use complement rather than work against each other.

This view of ourselves, however, assumes we are all fully rational, toting up the benefits and costs of everything before making the best choice to meet our goals. However, we know we do not act rationally all the time, and irrationality greatly affects our ability to make choices in the area of climate change.[32] Sometimes even if we know of the risks of a given action, we do not adequately account for these harms and so engage in the harmful act. We face all sorts of rationality problems, such as "motivated disbelief, the ostrich effect, confirmation bias, present-bias, adaptation, and intangibility."[33] Jaccard points to how "self-interest, delusion, and risk" are interconnected and are influenced by outside factors.[34] We worry more about risks that are salient (like a recent plane crash) than ones that we cannot readily bring to mind (like persistent future droughts). We do not understand small probabilities of large harms – which are at the centre of the problem of climate change. We are much more sensitive to losses than gains – and thus much less willing to risk losing out on what we already have through a policy change. We need to think about how to design policies to take account of these barriers to rational decision-making.

Some choices that appear irrational may not be. Dan Kahan at Yale Law School argues that people look to what others in groups they identify with believe about climate change. In the United States, you can think of Republicans versus Democrats, but the parallels are everywhere – Liberals versus Conservatives, Albertans versus Quebeckers, and the like. And Kahan says this is rational.[35] If your group tends to believe climate change is not a big deal, you are more likely to as well. Your chance of being harmed by climate change is low and your individual actions seem to be ineffective, but the risk of being estranged from your group for not believing the same things is large. You can be ostracized for challenging the norms. This is another instance where what is rational for an individual may not be best for society as a whole, just as if individuals sit back and allow others to bear the costs of reducing emissions nothing may get done, as everyone may free-ride in this way.

A further source of disconnect between the actions people know would help and those they actually take is the possibility that they lack the willpower. I may want to run a half-marathon with my daughter, but when it comes to actually putting in the time to train I may find excuses not to run. We may know our emissions and even the harm from those emissions, but we may not be able to commit ourselves to actually reducing them.

People do not generally just compare the cost of an energy-efficient washing machine with the benefits in terms of the energy saved and possibly the emissions reduced. Their calculations depend on how much weight they put on costs or benefits in the future as opposed to right now. We talked about this "discounting" in chapter 3. People may be willing to say they want to run in a race in three months' time or commit to giving a talk in the fall. However, as the time gets closer, their preferences seem to change. When you have to actually get up in the morning for the run or take the time to travel to where you have to give the talk, you do not want to anymore: the cost seems too high. In short, people often prefer smaller benefits today (buying the lower-cost washing machine) over larger rewards later (reaping lower energy bills). Delay means you value the larger future reward less, even though you will likely regret it.

Finally, we often assume our current choices are independent of our past ones. But we know that is not true. You may be aware of the health concerns about smoking and commit to stopping, but when the time comes, you smoke anyway – partly because it is addictive, but also because it can become ingrained in who you are and how you get through the day. The same is true of our choices around climate change. We drive rather than walk because it is easier, but also because it is something we have become accustomed to doing and have built our lives around. It is a habit.

Law, Institutions, and Norms

Our institutions and policy choices can impact norms and decision-making. What policies we want will depend on where we think the problem lies. Policies aimed solely at providing information will fail if people make mistakes even if given the information or if they do not believe what they are being told. Changing the price might not work or not work sufficiently well if there are strong non-price forces at work, such as norms of what type of car to drive or whether to drive.[36] Norms could in some cases be stronger than pricing.[37]

We can think of three broad categories of policies: those that give people better information or education, those that attempt to more directly change choices, and those that seek to involve people in policy choices.

Information and Norms

One possible explanation of why people do not change their behaviour is that they care about the environment and other people but they do not know enough. For example, people may not know what their emissions are, but if they did they would reduce them. Alternatively, perhaps they know their emissions but are less clear about the harms from those emissions.

If we believe individuals do not take climate action because they do not know enough, the obvious solution may seem to be to give them more information. In fact, this has been a key plank in Canadian climate policy, particularly in the early years. Policies were aimed at telling people about what climate change is and what they can do to help – and then left the choice of what to do up largely to them. Many of the federal climate plans going back to Conservative prime minister Brian Mulroney's 1990 Green Plan have proposed funding information and education campaigns.[38] In order to make climate policy politically acceptable, "three decades of Canadian climate policies have been dominated by non-compulsory information and subsidy programs."[39]

Some of this information informs individuals about their choices. Labels on cars, appliances, and even food (such as where it was grown) help individuals steer their purchasing decisions in more climate-friendly directions if they want. Other attempts have been made to give people a sense of their overall success at reducing emissions, such as calculators that allow people to determine their personal "carbon footprint" (how much carbon they emit in a year).

Related efforts have been made to steer decisions not by showing the energy or emissions-related information about a choice but the other benefits, such as to health. Eliminating the use of natural gas to heat your home can, for example, improve the air quality within your house for you and your children. Driving less and walking more can have significant health benefits. Framing of choices to emphasize the benefits such as to health can be more effective at altering decisions than focusing on the costs of not acting.[40]

Such information campaigns not only let people know if they can do more but also connect into social norms. As we saw, people may decide to reduce emissions if they feel that they will somehow be thought less of if they do not or if it makes them feel guilty to choose higher emissions options. Information can be tied into what others are doing. For example, telling people what their emissions are relative to others like them can lead to reductions in emissions, such as where energy companies try to encourage energy savings by showing a household's monthly usage and bill compared to the average for their neighbours.[41]

A related possibility is education – that is, start to build knowledge of climate and environmental issues through the education system (particularly at a young age) or other avenues for helping people understand, such as books or television programs.

Education can give people a basic understanding of the way climate works and the general outline of the science. They still have to make their own choices, but such an understanding may provide a common basis for discussion and a shared foundation for identifying reasonable choices.

Providing information and education will not be sufficient. As we saw, people tend to believe what and whom others in their group believe. And we may not be able to understand or make the connection between our emissions and the harms of climate change even if we are told. However, building trust in common sources of information and ideas may increase the probability of reaching common ground about what needs to be done.

Changing Choices

If we are to do more than just give people information or hope that norms develop, how forceful a role should government take? We have already seen that we have a range of tools that can be effective in reducing emissions. For industry, we can price emissions or we can regulate how different products are made or how power is generated. All these tools are necessary to make progress on climate change. Similar tools can be employed to reduce individuals' emissions and, as importantly, to influence norms.

The tools can be arrayed along a spectrum from those that allow individuals the most choice to those that allow the least choice. First, stemming from the idea that people are not rational and often do not choose options that are in their best interest, some policy-makers favour "nudges" – altering the way choices are presented to help people make better choices. Defaults are the choices or rules that apply if no one changes them. They "establish what happens if people do nothing at all."[42] Nudges preserve people's choice but influence them to choose what we think they would choose if they had the information and could process it. People, for example, may be more likely to save for retirement if they have to opt out of retirement plans rather than opt in.

Nudges in the climate context could make people have to actively choose to not use the less-emitting option. For example, in Germany many electricity providers automatically enroll customers in renewable energy programs.[43] Even small changes can make a difference. For example, one study found that a 1 degree C decrease in the default thermostat setting led to a reduction on average of 0.38 degrees C in temperature chosen.[44] "Green defaults" may work because defaults suggest the "right" choice, because people are lazy or procrastinate, or because defaults set the baseline from which people view change as a loss.[45]

However, we have to think carefully about changing defaults – they require that those setting the default know the choices that people would prefer if they were fully informed and completely rational. Governments may not always have this information. But defaults are generally not climate friendly, and changing them can make a difference in our choices. Such interventions "may not significantly address the more basic causes, or the magnitude, of contemporary policy problems" such as climate change,[46] but they can help.[47]

A second set of tools involves altering the price of choices people make. In one sense, information does this already. Giving people information about what happens if emissions do not go down tells them about the true costs of their actions. But governments can more explicitly change the price – either by increasing the price of choices that emit greenhouse gases or, equivalently, reducing the cost of taking more climate-friendly choices. We have already discussed carbon pricing, but governments have done more than impose a flat carbon price. Some, for example, vary electricity prices according to the time of day – such as increasing the price during periods of peak use or reducing the price during periods of low use (like the middle of the night).[48]

Similarly, governments have tried to subsidize certain decisions. Like information, subsidies have been central to our climate policies, in part because they are more politically acceptable than raising prices. For example, both the federal and provincial governments subsidize individuals buying electric vehicles.[49] The effectiveness of subsidies may be seen in the decrease in the number of electric vehicles purchased in Ontario when the provincial government cancelled its subsidy.[50] However, there are downsides, such as where the government tries to steer technological choices by "picking winners." Governments should frame subsidies broadly to allow performance to be the key – for example, subsidies for electric vehicles, or zero-emitting vehicles, and not for particular technologies (like those produced in Ontario). There is also an issue about the cost-effectiveness of subsidies. The Ecofiscal Commission, for example, found Quebec's program to subsidize electric vehicles was a very high-cost way to increase the use of electric vehicles.[51] Again, however, subsidies may be a less politically challenging method of reducing emissions than increasing an explicit carbon price.

One related policy area that might help bring change is public financing of infrastructure. For example, provincial governments are setting up charging stations for electric vehicles or subsidizing public transit. This form of public financing of infrastructure lessens the cost to individuals of reducing emissions and will be an important part of the path forward. It helps people more easily make the climate-friendly choices and does so on a large scale. Governments need to make it less costly and more attractive to live with less carbon.

Pricing still allows choice. You can continue to drive your car if you are willing to pay the added price of gas. However, price can be seen as unfair. Rich people do not have to cut back as much, or possibly at all. Moreover, changes in behaviour that come from a modest price change will take too long to make a difference. If we need immediate action, there is another option – regulate choices, such as prohibit driving in some contexts, limit the options for heating your home, or require use of electricity more generally. Some argue that because individuals are so bad at choosing, bans are necessary to actually make a difference.[52] Governments do regulate choices of individuals, for instance in the requirements around vehicle fuel economy or building standards. Of course, limiting choices can be politically difficult.

We therefore have a range of ways to change people's choices – give them information, change prices, and mandate certain choices. What connection do these approaches have to changing the social norms necessary for broader change, including political support for strong action? First, our laws and policies signal common understandings.[53] They can change how we view the world by expressing a common goal that provides a basis for particular choices. Laws on littering, for example, may not be well enforced, but they can reduce littering by expressing society's disapproval of throwing garbage on the ground. This sense of disapproval can provide a basis both for the belief that others will look down on us for littering and for a sense of guilt if we litter. One hope is that setting emission targets in law, imposing even a low carbon price, or requiring some emission reductions can change people's norms about what is appropriate. The information comes from the fact that there is a law.

Second, law and policies can help with commons problems not only by directly changing behaviour but also by giving people comfort that others will also bear costs. We have discussed how preserving a common resource may require a belief that others are also sacrificing. Climate laws and policies can aid in building that sense of common action, of the reciprocity needed to build common action. It can happen through the mere fact that there is a law or policy. It could also happen through the results of the laws or policies. For example, subsidies for electric cars may allow individuals not only to reduce their own emissions but to signal to others that reducing emissions is the right thing to do, and so possibly change their choices as well.[54]

Finally, changing behaviour or choices can have a long-term impact on strengthening any norms that currently exist. It is like a stream of water that deepens the channel as it flows. Having people adopt climate-friendly practices can strengthen their understanding and willingness to support more action. Economists talk about experience goods – things that you cannot really understand the benefits of until you use them. Climate action may be similar, since it is easy to overstate the costs of change on your lifestyle or underestimate the benefits such as to your health. Moreover, taking action that reduces emissions can become part of people's identity – who they are and how they view the world. Such changes can build over time, strengthening support for greater climate action.

Participating and Deliberating

One difficulty with information or defaults or even regulations and pricing is that they build on the notion that governments know what people want or should want. Are people making mistakes and so governments should help them do better – or is that too paternalistic?[55] Defaults or legislating choices, for example, work if governments can identify what "most people would choose if they were adequately informed."[56] But it is not entirely straightforward to figure out what that choice would be, or what to do if the default benefits many but not really the person who is subject to the default (that is, the person forced to choose is bearing a cost that others gain from).[57]

How we create our laws makes a difference in this regard. Laws can be imposed by government without participation by the public. Sen likens this approach to treating people like "motionless patients" or "passive recipients" for policy to operate on.[58] It enforces the idea that governments know what is best for people to do and they should just follow along. As we discussed last chapter, we have made some progress on ways to get people involved in policy decisions, although the path has not been straightforward. Some governments have seen public participation as too costly and too subject to mistakes and have reduced the amount of participation or its importance in the policy process. We have seen, for example, ebbs and flows in public participation in Canadian environmental assessment processes for large projects.[59]

Much of the Canadian policy process is based on notice and comment – governments tell Canadians what they are proposing and then allow them to send in comments. The transparency is good, but the process lacks opportunities for people to talk together to find common ground or to develop new norms about appropriate action. People need to deliberate about the best course of action. This deliberation among Canadians can lead to a common understanding of what needs to be done and hopefully greater buy-in on reductions.

Our institutional process and legal requirements on participation feed into this notion of "deliberative democracy" – that policy decisions need to be more than just an aggregation of what people currently seem to prefer. It has proven difficult to find broad-based forms of deliberation. However, what courts require of governments in terms of participation or legislation mandates for the processes of policymaking can bring people into the process or effectively exclude them.[60] For people to trust in and support policies to reduce emissions that can involve significant costs and change, they have to internalize the need to act – and that requires better means of involving them in decisions.

Crises, Institutions, and Norms

Economists often assume that individuals do not change their preferences – what they want. However, people do change. It is just not easy to get people to change, and the pathways are not always clear. When they do change, markets can change, as can political calculus. Aldo Leopold, a famous environmentalist from the 1950s, stated in *The Sand County Almanac* that "no important change in ethics was ever accomplished without an internal change in our intellectual emphasis, loyalties, affections and convictions … In our attempt to make conservation easy, we have made it trivial." Relying on subsidies and vague informational policies has undermined real transformation.

Long-term, consistent action requires such an internal change. There are signs from youth movements and statements by many business leaders and politicians that norms are changing, but it can take time. We have talked about how laws and institutions can help build norms. Another way people change is in response to a crisis – if things really, visibly fall apart then people can adjust quickly. We have seen tremendous change in how individuals interact because of the pandemic. Social distancing and working from home quickly became part of everyday life. Many businesses altered processes to help make masks or sanitizers. Individuals volunteered their time and resources to help in the fight.[61] We are seeing the portends of the crisis stage for climate change with increased floods, storms, and wildfires, not to mention rising sea waters and a constant barrage of stories about droughts and disease.

However, the pandemic has also shown us limits on such change. There was tremendous rapid reaction because of fear of immediate harm from the coronavirus.[62] Once this initial reaction wore off, some of the changes in norms and behaviour weakened in the face of fears over the economy. The United States and to some extent Canada also saw tremendous backlash once responding to the pandemic became polarized politically. Norm change in the climate context may be slower not only because of the lack of evident immediate harm, but also because it is similarly subject to these obstacles from economic self-interest and politicization.

Hopefully we can act before things get worse.

chapter twelve

Breaking the Cycle

Canada has failed to adequately address climate change. We know a lot about climate change and how to address it; we can see the core of the problem, from both the science and social science side. We know what we should aim at, at least as a minimum – to keep temperature increases below 1.5 degrees C. We also have a sense of the underlying problem – we are mired in a series of commons problems as individuals, as provinces, as countries.

Fortunately, we also have a detailed understanding of what tools might be successful at reducing emissions and at what cost. We can use carbon prices or regulations or subsidies to get individuals and industries to reduce emissions. We can invest in "green" infrastructure in the form of increased public transit or charging stations for electric vehicles or more bike lanes. They will all help, though some are more expensive than others. The Canadian Institute for Climate Choices has come out with a list of "safe bets" and "wild cards" that show a range of actions Canadian governments can take.[1]

But change is not happening fast enough. We have discussed problems in all three key elements for effective policy: ideas, interests, and institutions. We have too thin a vision of what we should be aiming at. We have focused on economic growth as an end in itself, with the environment and issues of equality as at best constraints on growth. We seem more recently, at least in rhetoric, to recognize the possibility of having both economic growth and environmental improvement. While this is progress, it is too weak a foundation to support the structure needed to bring about real change. We need to build in elements that allow all Canadians to live lives they have reason to value.

There are also powerful interests weakening the footings of change. We have talked about how Canada has grown through reliance on its natural resources – fur, wheat, fish, forests, minerals and, of course, fossil fuels. Environmental law, and climate change law in particular, has been hindered by economic interests – companies that want to expand but also workers that see their futures tied to particular industries and people with pensions who want a secure future. These interests, while understandable, make change more difficult. Moreover, as with other countries, we have become accustomed to living with all that fossil fuels bring – easy transportation, inexpensive heating, cheaper goods and food.[2]

Even as we have grown aware of the weaknesses of this approach, and of the coming collapse due to climate change, Canadian institutions have been unable to bring about real change. Politicians unsurprisingly respond to the interests resisting change. It is hard to lead when there are few who see the need to follow. But the problem is greater than that. Canadian institutions have not been able to effectively mediate between competing interests resisting and promoting change. Governments make few commitments, and even those that are made are undermined by opaque exemptions and buyouts. The courts seem reluctant to hold governments to what weak commitments they have made.

Collectively, our ideas, interests, and institutions have all provided an unstable footing for effective action. In a sense, the issue is less a "climate change" problem than a governance problem. We see governments making big promises for change – going to net-zero by 2050, eliminating coal use, putting higher prices on carbon. The news is replete with stories of how the markets are transforming and new sources of energy are being developed. Yet worries remain. Will the changes happen, or happen fast enough? We need urgent action but we have not seen it so far. The more we delay, the more damage there is to people and the environment and the higher the cost of future change.

Moreover, even if change occurs, it can happen in many different ways. Canada can get to net-zero in ways that are unfair to others – to other countries, across regions in Canada, and across individuals. We have seen as much in the pandemic. Much of the harm was visited on vulnerable groups, but many of the solutions were only partially aimed at helping them – either those in poorer countries or those in Canada. The same is true for climate change. We have talked about Baptist and Bootlegger coalitions; one thing to remember about the bootleggers is that you are leveraging their interest, not their principles. We need to take advantage of their innovation and drive, but at the same time we need to think about where it is taking us.

Canada may be embarked on a new wave of environmentalism, spurring action on climate change. How do we keep the wave from crashing? We need to go back to why such waves have foundered in the past.

The Canadian Problem

Much of what we talk about in terms of climate change solutions is not particular to Canada. The tools necessary to reduce emissions – pricing, regulation, public investment, private markets – are common across countries. Canadian governments rely too heavily on the notion of economic growth and often pay short shrift to the environment, Indigenous peoples, or others who bear the costs of growth – but so do other countries, to a greater or lesser extent.

Meaningful change, however, requires a better understanding of Canadian institutions and where they came from. Institutions are the wedge between what theory tells us are the best solutions and how we can actually act. Canada's story has been about reliance on various staples, starting with fish and fur and moving on to forests, minerals, and fuel. These resources fueled Canadian economic growth, but its reliance on them also created problems. Canada has exploited one resource and then seem to move blithely on to another – either willingly or because the resource was used up or severely depleted.

As we discussed, a big source of this pattern is that Canada built its laws and institutions around exploiting resources – to the detriment of the environment, Indigenous peoples, and others.[3] This foundation has led to weak environmental laws generally.[4] It has been particularly problematic in finding the will to act on climate change. Canadian laws are replete with discretion, giving flexibility and the promise of decisions based on expertise and knowledge but also providing large scope for decision-makers to be swayed from the public interest by powerful resource-industry interests. Responsibility is diffused across different levels of governments, allowing them to blame each other but take little action. The courts and public have tended to defer to government decision-makers, providing only a limited check on the misuse of discretion and on the finger-pointing.

This all creates a particular Canadian vicious circle.[5] The public becomes upset about some environmental issue (likely when the economy is going well), raising some pressure for government action. Each government points at other governments as being primarily responsible, deflecting blame. When they do act, they tend to enact progressive-sounding legislation that delegates broad discretion to various government actors. Industry has had input both in the legislation itself and even more so in how the delegated discretion is exercised. In addition there are others who, concerned about the economy or jobs for example, also seek to weaken the application of the laws. If there is little action on the issue, the courts have tended to defer to the choices of government actors, and the public will often move on to other concerns. Because the courts defer to elected officials and the public do not pay adequate attention, those officials have space to further weaken environmental laws. Eventually the same or a different problem will arise, and we start all over again.

Creating a New Circle

This pattern creates the waves we have seen in environmental law and has fueled the waves in climate law and policy. But the situation is not all bad. Canadian institutions also provide a basis for change. We have discussed climate change as a commons problem at many levels, from global to national/provincial down to individuals. Some see the commons problem as the wrong analogy. Political scientists Matt Hoffman and Steven Bernstein prefer the notion of a "fractal carbon trap."[6] We are locked in politically, economically, technologically, and culturally to the use of fossil fuels at all levels from local to global. The dependence at each level reinforces the lock-in at others. Hoffman and Bernstein point to the need to get over a threshold of action to escape the trap; otherwise we are inevitably drawn back to fossil fuels. If we pass this threshold, however, change towards decarbonization will be self-enforcing and ultimately successful.

This view is similar to our discussion of carbon as the new "staple." Canada built its society and institutions around fossil fuel use, which limits the ability to change and even, as legal scholar Jason MacLean puts it, to imagine what change is possible. Hoffman and Bernstein highlight how this trap can emerge largely without a central plan, while others see a more active role from industries that benefit from the lock-in. In this book I argue that Canadian institutions – the constitutional structure that fosters cooperation, the commitment to reasoned decision-making, the history of sharing risks – provide the foundation from which to build a new cycle.

We have focused in this book on three broad steps. First, we need to reduce opposition to change. Opposition comes from those who fear that they will lose out in the transition, that the costs of change will be visited on them while others benefit. Workers in industries dependent on fossil fuels worry that they and their communities will be left behind. Vulnerable individuals worry about the costs of the new "green" economy. Yet the history of Canada has been one of finding ways to share risks. Out of the Depression came the beginnings of a strong welfare state that reduced inequality and risks from unemployment, sickness, and aging. The programs have not been perfect by any means, but the move to support the whole individual, and not just hope economic growth will solve all problems, provides the foundation for addressing fears of the transition. One area important to a "just transition" in which we are woefully lacking is our relationship with Indigenous peoples, addressing which will need to be part of the path forward.

To the extent that it is regions that worry about bearing the brunt of change, Canada also has considerable experience in sharing risks across provinces. Most provinces have good times and bad times, so the provinces that are doing well at a particular time support those that are not. The federal equalization, stabilization, and transfer programs provide the basis for developing transfer programs to share costs of reducing emissions across regions. Changes to these programs, or a new conditional grant program, can ease the costs of change for provincial governments. In addition, we know which industries are particularly at risk from the transition, which is closely tied to which regions will bear more costs. Government transition policies need to help only those industries that need it and only when they need it. Transition policies must be reconsidered so their costs – subsidies or greater emissions – are used in the most efficient and fair manner possible.

Second, we need to build trust to support better decisions. The diffusion of responsibility across levels of government has led in many cases to less environmental protection, particularly where the federal government refuses to act. However, it does provide the Canadian system with the ability to tailor rules and to experiment. The constitutional principle of cooperation helps find the balance between the need for national action and the value of local decision-making. As we saw, Canadian governments have failed to employ this notion of cooperation in a self-reinforcing manner, but instead have sought consensus, which has tended to lead to weak environmental rules. Building the trust necessary for the federal government to support the transition and for provincial governments to feel that they are not being taken advantage of is paramount.

Trust is also central to how each government makes its decisions. We lean too heavily on political accountability and discretion without any real support. All three key supports for policies need to be built up: expertise, political accountability, and legality. Climate legislation will of necessity involve grants of discretion given the scope of the changes required. Setting targets and specifying factors that need to be considered will help hold decision-makers to account but are not on their own sufficient. We need to strengthen the role of independent expert decision-makers in policy decisions. Climate decisions involve value choices, so experts cannot have the final say, but effective policy requires more than experts offering advice that the decision-makers can ignore. Further, the courts need to play a greater role in holding government decision-makers to their task – requiring that they justify their decisions clearly and show how they are furthering the legislative objectives. Each change builds on the others, increasing the opportunity to get the best out of discretionary powers and limiting their downsides.

Finally, we need to strengthen interests that push for change. Both the federal and provincial governments have begun pouring money into the economy in an attempt to spur growth in the new "healthy" economy. They have also sought ways to foster innovation, such as through new regulatory frameworks and through encouraging private investment. As the new economy grows so too can support for change. Political scientists Jeff Colgan, Jessica Green, and Thomas Hale note that as investments change from fossil fuels to other forms of energy, the politics can change.[7] Changes can be "self-reinforcing," as these new industries will have resources to lobby for change and politicians will become reluctant to put in place rules that harm them.

But we also need to consider the role individuals can play in supporting change. We spoke of reducing fear, but we also need to build the belief in the need for action. By this I mean not belief in a particular outcome but an understanding of the dilemma Canada is facing and what the actual choices are. As we discussed in the last chapter, social norms underlie our choices both as consumers and as voters. Some change may come as people become accustomed to life with less carbon, realizing that the costs in their lifestyles are not as great as imagined and the benefits greater. Further, social norms may grow through the experience of harms from climate change. However, we can do more. Our policies can provide information and education, but how laws and policies are developed also needs to change so that people can come to feel like drivers in the process as opposed to passengers. We need to build towards a common view of the path forward to ground more urgent, fairer action.

The New Circle

These elements – reducing opposition, building trust, strengthening support – are all interwoven and support each other. The politically accountable legislature gives clear powers and mandates to government actors. These government actors use independent expertise and consultation with the public to ground reasons for how they use these powers. The courts require government reasons to be transparent and intelligible and to justify the decisions. The public gains trust in the decisions of the legislature and the government from this process, which both helps build norms of climate action and provides the public with the basis for holding governments to account. Everyone has a role to play, and each is involved in the process of building and supporting an effective climate plan.

And trust across levels is self-supporting. Trust between countries is built when countries make and keep commitments to act. Trust and cooperation within a country grows with a system that respects and builds on the powers and interests of different regions and shares the risks of action. Trust in government arises from setting reason-based targets and having a system that builds on them, and by knowing that other provinces and countries will act.

Of course, such a self-sustaining ecosystem is hard to put in place and can easily get out of balance. Governments may be reluctant to make stringent promises about climate change if they believe they will be held to them by the courts. The political risk is high if the public is not yet completely on board and other countries seem reluctant to act. Even if stringent promises are made, they are likely to be watered down as they are applied.

Many of the accounts of what we do now seem to ignore the dynamics at play and show us a set of seemingly independent end points – strict targets, efficient tools, more transparent government, independent experts, and voting for legislators who will make this all possible. But how do we set all this in motion?

Getting buy-in for strong climate action requires a plan based on sharing of risk. This can overcome the strong interests resisting change. How do we get such a plan? New directions can come from a crisis. The Great Depression brought home the widespread risk to individuals from underregulated markets. Fighting a world war gave rise to a civic spirit to fight an external foe. The COVID pandemic brought a realization of a shared vulnerability. A risk that hits everyone can bring about change – a social safety net, acceptance of rationing, or willingness to self-isolate and fund support programs. It is of course easy to romanticize these changes. Each of these events and the response to them brought greater risks and harms for some than others. Climate crises can, and are, bringing about some such change, but not quickly enough and not always in a considered, fair manner.

The key to rapid, significant change returns us to how we began this book – with Bertrand Russell's advice to future generations. His first piece of advice was to look at the facts, not what you wish to believe or what you think would have good effects. We have good information about the threats from climate change as well as the potential costs of both acting and not acting. We know the high costs that are being imposed on vulnerable individuals here and abroad and the consequences to future generations of failing to act. We also have considerable well-researched guidance about how to reduce emission and change our energy systems. It is clear we need to act and that action will be difficult.

The second part of the Bertrand Russell quote, though, is as or more important: "love is wise, hatred is foolish." We need to recognize that change comes with real fears and real losses. We need to understand that people come at this issue from very different places, and the only way we are going to find a path forward is to find empathy and a willingness to listen. We are all in muddy waters and it is hard to see where to move, but we should use that recognition of mutual uncertainty to base a common approach, to find empathy for others in the same waters.

The matter thus in part comes back to values and choices. We know people will choose differently in response to changes in price – you lower the cost of electric cars and people will buy them, raise the price of gas and people will drive less. However, individuals also alter their choices through changes in their underlying values.[8] These values are the foundation for real action, on which all other changes hinge. We need an appeal not only to people's interests but to their deeper values.

Economist Albert O. Hirschman pointed to the need to pay attention to love and civic spirit as well as interests. As he notes, economists have tended to argue that "the social order is more secure when it is built on interest rather than on love or benevolence."[9] He by contrast views love, benevolence, or public spiritedness as a resource "whose supply may well increase rather than decrease through use" and that becomes weaker if not used. We come to believe trust and public spiritedness are not necessary for the functioning of an economic and political system, and so self-interest takes over. However, they *are* necessary.

But the interaction is not straightforward – you cannot just rely on love and public spiritedness, as "they atrophy when not adequately practiced and appealed to by the ruling socioeconomic regime, yet will once again make themselves scarce when preached and relied on to excess." And so self-interest and public spiritedness go back and forth – at times of crisis you can rely on trust and love to make progress, but they then give way to self-interest for a period, which again gives way to public-spiritedness when it is clear self-interest is too thin a base on which to stand.

We need to find a balance – build up notions of shared risk and trust to ground climate policies and not merely rely on self-interest. We need to bring what Hirschman calls love and public-spiritedness to the centre of our approach so it can grow and allow for stronger climate action. Similarly, trust builds on itself – it gets stronger as we build reasons to trust and weaker when we abuse it or ignore it. Building this trust through our institutions is central to the path forward.

Luckily, we have the basis from which to move. What some argue are obstacles to climate policy – the Canadian federal structure and Constitution, the equalization and stabilization programs, the open economy –are the key building blocks. We can use the basic risk-sharing framework of the Constitutional structure and the social safety net policies to bring individuals and provinces to support decarbonization. We can use the basic Canadian legal framework that emphasizes reasons and reason-giving to support better decisions and greater accountability and hence trust. We have elements of a successful system to overcome the powerful gravity of the status quo. Many individuals are worried about change; we need to recognize and accommodate those worries in order to move forward. We have shifted in recent years towards markets as the solution, but it is time to shift back to a greater emphasis on policies that promote sharing of risks and opportunities.

So sharing, trust, love – for me, these form the ground from which we can grow. As I said at the beginning, it seems both too little and too much. It lacks the hard edge that policymakers like to show off. At the same time it is asking a lot. We should aim high.

Notes

Preface

1 Kathryn Harrison, *Passing the Buck: Federalism and Canadian Environmental Policy* (Vancouver: University of British Columbia Press, 1996), 6.

2 Parliamentary Budget Officer, *Scenario Analysis: COVID-19 Pandemic and Oil Price Shocks* (Ottawa: PBO, 2020), https://www.pbo-dpb.gc.ca/web/default/files/Documents/Reports/RP-192 0–033-S/RP-1920–033-S_en.pdf.

3 See, for example, Bill Weir, "What Coronavirus Could Teach Us about Climate Change," CNN, 26 March 2020, https://www.cnn.com/videos/weather/2020/03/26/weir-climate-crisis-impact-coronavirus-project-planet-orig.cnn/video/playlists/project-planet/; Cameron Hepburn et al., "Will COVID-19 Fiscal Recovery Packages Accelerate or Retard Progress on Climate Change?" *Oxford Review of Economic Policy* 36, no. S1 (2020): 5, https://doi.org/10.1093/oxrep/graa015.
Weir, "What Coronavirus Could Teach Us."

4 In addition, there are connections between the pandemic and climate change, with the pandemic making certain already-existing issues worse, such as food crisis, and climate change increasing the risk of pandemics. See, for example, Smart Prosperity Institute, *The Resilient Recovery Framework: Version 3* (Ottawa: Smart Prosperity Institute, 2020), 11, https://institute.smartprosperity.ca/sites/default/files/resilient-recovery-framework-v2.pdf.

5 Guido Calabresi and A. Douglas Melamed, "Property Rules, Liability Rules, and Inalienability: One View of the Cathedral," *Harvard Law Review* 85, no. 6 (1972): 1089–128, https://doi.org/10.2307/1340059.

6 Calabresi and Melamed, "Property Rules," 1128.

1 Of Fear and Loathing in Canadian Climate Policy

1 Z.A. Wendling, J.W. Emerson, A. de Sherbinin, D.C. Esty, et al. *2020 Environmental Performance Index* (New Haven: Yale Center for Environmental Law & Policy, 2020), epi.yale.edu.
2 Myles Allen et al., "Summary for Policymakers," in *Global Warming of 1.5°C*, ed. Valérie Masson-Delmotte et al. (Geneva: Intergovernmental Panel on Climate Change, 2018, 2019), 5, https://www.ipcc.ch/site/assets/uploads/sites/2/2019/06/SR15_Full_Report_High_Res.pdf.
3 Council of Canadian Academies, *Canada's Top Climate Change Risks: The Expert Panel on Climate Change Risks and Adaptation Potential* (Ottawa: CCA, 2019). This work, https://cca-reports.ca/wp-content/uploads/2019/07/Report-Canada-top-climate-change-risks.pdf, discusses the potential impacts in Canada of climate change.
4 F. Warren and N. Lulham, eds., *Canada in a Changing Climate: National Issues Report* (Ottawa: Government of Canada, 2021), https://changingclimate.ca/national-issues/.
5 For example, both Amazon and General Motors announced that they intend to be "carbon neutral" by 2040. See David Welch, "GM to Only Sell Zero-Emissions Models by 2035," *Fortune*, 29 January 2021, https://fortune.com/2021/01/29/gm-net-zero-vehicles-2035/; Ken Silverstein, "Amazon's 'Restless Nights' Center on Achieving Net-Zero by 2040," *Forbes*, 18 February 2021, https://www.forbes.com/sites/kensilverstein/2021/02/18/amazons-restless-nights-center-on-achieving-net-zero-by-2040/?sh=3757c3646dac.
6 United Nations Environment Program, *Emissions Gap Report 2021* (Nairobi: United Nations, 2021).
7 Bill C-12, Canadian Net-Zero Emissions Accountability Act (Royal Assent 29 June 2021).
8 Josh Lederman and Denise Chow, "Biden Commits to Cutting U.S. Emissions in Half by 2030 as Part of Paris Climate Pact," NBC News, 22 April 2021, https://www.nbcnews.com/politics/white-house/biden-will-commit-halving-u-s-emissions-2030-part-paris-n1264892.
9 Ian Austen and Christopher Flavelle, "Trudeau Was a Global Climate Hero. Now Canada Risks Falling Behind," *New York Times*, 21 April 2021, https://www.nytimes.com/2021/04/21/world/canada/trudeau-climate-oil-sands.html.

10 S. Langlois-Bertrand, K. Vaillancourt., L. Beaumier, M. Pied., O. Bahn, and N. Mousseau, *Canadian Energy Outlook 2021 — Horizon 2060* (Montreal: Institut de l'énergie Trottier and e3c Hub, 2021).

11 Kathy Harrison, *Passing the Buck: Federalism and Canadian Environmental Policy* (Vancouver: University of British Columbia Press, 1996), 6.

12 Mark Jaccard, *The Citizen's Guide to Climate Success: Overcoming Myths That Hinder Progress* (Cambridge: Cambridge University Press, 2020).

13 See, for example, Steven Croley, "Theories of Regulation: Incorporating the Administrative Process," *Columbia Law Review* 98, no. 1 (1998): 23. In the environmental context, see David Markell, "'Slack' in the Administrative State and Its Implications for Governance: The Issue of Accountability," *Oregon Law Review* 83 (2005): 1.

2 Cows, Cod, and Coal: The Roots of Canada's Climate Dilemma

1 Mark Jaccard, *The Citizen's Guide to Climate Success: Overcoming Myths That Hinder Progress* (Cambridge: Cambridge University Press, 2020), 43–7; Spencer R. Weart, *The Discovery of Global Warming*, rev. and expanded ed. (Cambridge, MA: Harvard University Press, 2008); Gale E. Christianson, *Greenhouse: The 200-Year Story of Global Warming* (New York: Walker and Company, 1999).

2 NASA, "2020 Tied for the Warmest Year on Record, NASA Analysis Shows," 14 January 2021, https://www.giss.nasa.gov/research /news/20210114/.

3 Elizabeth Bush et al., "Chapter 2: Understanding Observed Global Climate Change," in *Canada's Changing Climate Report*, ed. Elizabeth Bush and Donald S. Lemmen (Ottawa: Government of Canada, 2019), 29, 32, https://www.nrcan.gc.ca/sites/www.nrcan.gc.ca/files/energy/Climate -change/pdf/CCCR_FULLREPORT-EN-FINAL.pdf.

4 Elizabeth Bush and Donald S. Lemmen, eds., *Canada's Changing Climate Report* (Ottawa: Government of Canada, 2019), 5; Greg Flato et al., "Modelling Future Climate Change," in Bush and Lemmen, *Canada's Changing Climate Report*, 84.

5 Flato et al., "Modelling Future Climate Change," 84.

6 Christopher B. Field et al., "Technical Summary," in *Climate Change 2014: Impacts, Adaptation, and Vulnerability: Summaries, Frequently Asked Questions, and Cross-Chapter Boxes: A Working Group II Contribution to*

the Fifth Assessment Report on the Intergovernmental Panel on Climate Change, ed. Christopher B. Field et al. (Geneva: World Meteorological Organization, 2014), 62, https://www.ipcc.ch/site/assets/uploads /2018/03/WGIIAR5-IntegrationBrochure_FINAL-1.pdf.

7 Bush et al., "Understanding Observed Global Climate Change," 30.

8 Myles Allen et al., "Summary for Policymakers" in *Global Warming of 1.5°C*, ed. Valérie Masson-Delmotte et al. (Geneva: Intergovernmental Panel on Climate Change, 2018, 2019), 5, https://www.ipcc.ch/site /assets/uploads/sites/2/2019/06/SR15_Full_Report_High_Res.pdf; Tariq Banuri et al., "Drivers of Environmental Change," in *Global Environment Outlook GEO-6: Healthy Planet, Healthy People*, ed. Paul Ekins, Joyeeta Gupta, and Pierre Boileau (Cambridge: Cambridge University Press, 2019), 43, https://wedocs.unep.org/bitstream /handle/20.500.11822/27539/GEO6_2019.pdf.

9 Flato et al., "Modelling Future Climate Change," 77.

10 For the 1.5 degree resolution, see UN, Glasgow Climate Pact (Decision -/CP.26), 2021, https://unfccc.int/sites/default/files/resource/cop26 _auv_2f_cover_decision.pdf.

11 IEA, *Global Energy & CO2 Status Report 2019: The Latest Trends in Energy and Emissions in 2018* (Paris: International Energy Agency, 2019), "Coal," https://www.iea.org/reports/global-energy-co2-status-report-2019.

12 Lauren Sommer, "Carbon Emissions Are Falling, But Still Not Enough, Scientists Say," NPR, 14 April 2020, https://www.npr.org/sections /coronavirus-live-updates/2020/04/14/834295861/carbon-emissions -are-falling-but-still-not-enough-scientists-say.

13 IEA, *Global Energy Review 2021* (Paris: International Energy Agency, 2021), overview, https://www.iea.org/reports/global-energy-review-2021.

14 IEA, *Global Energy Review 2021*, overview.

15 Climate Transparency, *Brown to Green: The G20 Transition Towards a Net-Zero Emissions Economy: Canada* (Berlin: Climate Transparency, 2019), https://www.climate-transparency.org/wp-content/uploads/2019/11 /B2G_2019_Canada.pdf.

16 Allen et al., "Summary for Policymakers," 9.

17 US Global Change Research Program, *Fourth National Climate Assessment, Volume II: Impacts, Risks, and Adaptation in the United States, Report-in-Brief*, ed. D.R. Reidmiller et al. (Washington, DC: US Government Publishing

Office, 2018), 12–19, https://nca2018.globalchange.gov/downloads/NCA4_Report-in-Brief.pdf.

18 Ove Hoegh-Guldberg et al., "Impacts of 1.5°C of Global Warming on Natural and Human Systems," in *Global Warming of 1.5°C*, ed. Valérie Masson-Delmotte et al. (Geneva: Intergovernmental Panel on Climate Change, 2018, 2019), 245, https://www.ipcc.ch/site/assets/uploads/sites/2/2019/06/SR15_Chapter3_Low_Res.pdf; Library of Parliament, *Climate Change: Its Impact and Policy Implications* (Ottawa: Library of Parliament, 2020), 8, https://lop.parl.ca/staticfiles/PublicWebsite/Home/ResearchPublications/BackgroundPapers/PDF/2019-46-e.pdf.

19 Sandra Díaz et al., *The Global Assessment Report on Biodiversity and Ecosystem Services: Summary for Policymakers* (Bonn: Intergovernmental Science-Policy Platform on Biodiversity and Ecosystem Services Secretariat, 2019), 11–12, https://ipbes.net/sites/default/files/2020-02/ipbes_global_assessment_report_summary_for_policymakers_en.pdf.

20 International Labour Organization, *Working on a Warmer Planet: The Impact of Heat Stress on Labour Productivity and Decent Work* (Geneva: ILO, 2019), 13, https://www.ilo.org/wcmsp5/groups/public/-dgreports/-dcomm/-publ/documents/publication/wcms_711919.pdf.

21 Philip Alston and UN Human Rights Council, *Climate Change and Poverty: Report of the Special Rapporteur on Extreme Poverty and Human Rights*, A/HRC/41/39, 17 July 2019, 4, https://digitallibrary.un.org/record/3810720?ln=en; UN Department of Economic and Social Affairs, *The Sustainable Development Goals Report 2019* (New York: UN, 2019), 23, https://unstats.un.org/sdgs/report/2019/The-Sustainable-Development-Goals-Report-2019.pdf.

22 UN Human Rights Council, *Analytical Study on Gender-Responsive Climate Action for the Full and Effective Enjoyment of the Rights of Women: Report of the Office of the United Nations High Commissioner for Human Rights*, A/HRC/41/26, 1 May 2019, 3, https://digitallibrary.un.org/record/3807177/files/A_HRC_41_26-EN.pdf.

23 Alston and UN Human Rights Council, *Climate Change and Poverty*, 10. See also Miguel Molico, "Researching the Economic Impacts of Climate Change," Bank of Canada, 19 November 2019, https://www.bankofcanada.ca/2019/11/researching-economic-impacts-climate-change/; Matthew E. Kahn et al., "Long-Term Macroeconomic Effects

of Climate Change: A Cross-Country Analysis," IMF Working Paper WP/19/215, International Monetary Fund, Washington, DC, October 2019, 7, https://www.imf.org/en/Publications/WP/Issues/2019/10/11/Long-Term-Macroeconomic-Effects-of-Climate-Change-A-Cross-Country-Analysis-48691.

24 US Global Change Research Program, *Report-in-Brief*, 13.

25 Ove Hoegh-Guldberg et al., "Impacts of 1.5°C of Global Warming," 256.

26 Canadian Institute for Climate Choices, *Sink or Swim: Transforming Canada's Economy for a Global Low-Carbon Future* (Ottawa: CICC, 2021), 6, https://climatechoices.ca/wp-content/uploads/2021/10/CICC-Sink-or-Swim-English-Final-High-Res.pdf.

27 Bush and Lemmen, *Canada's Changing Climate Report*, 5.

28 Council of Canadian Academies, *Canada's Top Climate Change Risks: The Expert Panel on Climate Change Risks and Adaptation Potential* (Ottawa: Council of Canadian Academies, 2019), https://cca-reports.ca/wp-content/uploads/2019/07/Report-Canada-top-climate-change-risks.pdf.

29 Council of Canadian Academies, *Top Climate Change Risks*, 15.

30 F. Warren and N. Lulham, eds., *Canada in a Changing Climate: National Issues Report* (Ottawa: Government of Canada, 2021), https://changingclimate.ca/national-issues/chapter/overview/.

31 Canadian Institute for Climate Choices, *Tip of the Iceberg: Navigating the Known and Unknown Costs of Climate Change in Canada* (Ottawa: CICC, 2020), i, https://climatechoices.ca/reports/tip-of-the-iceberg/.

32 Canadian Institute for Climate Choices, *Tip of the Iceberg*, iv.

33 Allen et al., "Summary for Policymakers," 12.

34 World Meteorological Organization, *WMO Statement on the State of the Global Climate in 2018*, WMO-No. 1233 (Geneva: WMO, 2019), 4, https://library.wmo.int/doc_num.php?explnum_id=5789.

35 UNEP, *Emissions Gap Report 2020* (Nairobi: UN Environment Programme, 2020), xvii, https://www.unep.org/emissions-gap-report-2020.

36 UNEP, *Emissions Gap Report 2020*, xxi.

37 UNEP, *Global Environment Outlook GEO-6: Summary for Policymakers* (Cambridge: Cambridge University Press, 2019), 4, https://doi.org/10.1017/9781108639217.

38 Petteri Taalas and Joyce Msuya, foreword to *Global Warming of 1.5°C*, ed. Masson-Delmotte et al., vi.

39 Auditor General of Canada, *Reports of the Commissioner of the Environment and Sustainable Development to the Parliament of Canada: Fall 2017: Report 1: Progress on Reducing Greenhouse Gases: Environment and Climate Change Canada: Independent Auditor's Report* (Ottawa: Auditor General of Canada, 2017), 1, http://publications.gc.ca/collections/collection_2017/bvg-oag /FA1-26-2017-1-1-eng.pdf.

40 Nicole Dusyk et al., *All Hands on Deck: An Assessment of Provincial, Territorial and Federal Readiness to Deliver a Safe Climate* (Pembina Institute, July 2021), 13.

41 United Nations Framework Convention on Climate Change, art. 2, 9 May 1992, 1771 UNTS 107, https://unfccc.int/sites/default/files/convention _text_with_annexes_english_for_posting.pdf. The number of countries that eventually signed on to the UNFCCC has increased to 197.

42 ECCC, *Canada's Fourth Biennial Report of Climate Change* (Gatineau: ECCC, 2019), https://unfccc.int/sites/default/files/resource/br4_final_en.pdf.

43 Auditor General of Canada, *Progress on Reducing Greenhouse Gases*, 4.

44 See Auditor General of Canada, *Progress on Reducing Greenhouse Gases*, 13, for a list of actions taken up to 2017 by the federal government on climate change.

45 Government of Canada, "Greenhouse Gas and Air Pollutant Emissions Projections: 2019," last modified 3 January 2020, https://www.canada .ca/en/environment-climate-change/services/climate-change /greenhouse-gas-emissions/projections/2019.html.

46 UNEP, *Emissions Gap Report 2019*, xvi–xvii.

47 Auditor General of Canada, *Perspectives on Climate Change Action in Canada: A Collaborative Report from Auditors General* (Ottawa: Auditor General of Canada, 2018), 4, http://publications.gc.ca/collections /collection_2018/bvg-oag/FA3-137-2018-eng.pdf.

48 *References re Greenhouse Gas Pollution Pricing Act*, 2021 SCC 11.

49 Dusyk et al., *All Hands on Deck*, at 9.

50 Environment and Climate Change Canada (hereafter cited as ECCC), *A Healthy Environment and a Healthy Economy* (Gatineau: ECCC, 2020), https://www.canada.ca/content/dam/eccc/documents/pdf/climate -change/climate-plan/healthy_environment_healthy_economy_plan.pdf.

51 S. Langlois-Bertrand, K. Vaillancourt., L. Beaumier, M. Pied., O. Bahn, and. N. Mousseau, *Canadian Energy Outlook 2021 – Horizon 2060* (Montreal: Institut de l'énergie Trottier and e3c Hub, 2021), 92.

52 Dusyk et al., *All Hands on Deck*, 8.
53 Garrett Hardin, "The Tragedy of the Commons," *Science* 162, no. 3859 (1968): 1243–8.
54 Hardin, "Tragedy of the Commons," 1244.
55 See, for example, Elinor Ostrom, "A Polycentric Approach for Coping with Climate Change," *Annals of Economics and Finance* 15, no. 1 (2014): 97–134, ProQuest; William Nordhaus, "Climate Clubs: Overcoming Free-Riding in International Climate Policy," *American Economic Review* 105, no. 4 (April 2015): 1339–70; Scott Barrett, *Why Cooperate? The Incentive to Supply Global Public Goods* (Oxford: Oxford University Press, 2007).
56 Kelly Levin et al., "Overcoming the Tragedy of Super Wicked Problems: Constraining Our Future Selves to Ameliorate Global Climate Change," *Policy Sciences* 45, no. 2 (June 2012): 123.
57 Barrett, *Why Cooperate?*
58 In 2014 former Conservative prime minister Stephen Harper stated, "Under the current circumstances of the oil and gas sector, it would be crazy – it would be crazy economic policy to do unilateral penalties on that sector; we're clearly not going to do that." Margot McDermid, "Stephen Harper Says Oil and Gas Regulations Now Would Be 'Crazy,'" CBC News, 10 December 2014, http://www.cbc.ca/news /politics/stephen-harper-says-oil-and-gas-regulations-now-would-be -crazy-1.2866306. Similarly, Liberal prime minister Justin Trudeau noted that "there isn't a country in the world that would find billions of barrels of oil and leave it in the ground while there is a market for it." Mike De Souza, "Trudeau Approves Kinder Morgan Pipeline, Rejects One of Two Enbridge Projects," *National Observer*, 29 November 2016, http://www .nationalobserver.com/2016/11/29/news/breaking-trudeau-approves -kinder-morgan-pipeline-rejects-one-two-enbridge-projects.
59 There is an extensive literature on the ethics of climate change. See, for example, Ian H. Rowlands, "International Fairness and Justice in Addressing Global Climate Change," *Environmental Politics* 6, no. 3 (1997): 1–30; Eileen Claussen and Lisa McNeilly, *Equity and Global Climate Change: The Complex Elements of Global Fairness* (Arlington: Pew Center on Global Climate Change, 1998); Denis G. Arnold, ed., *The Ethics of Global Climate Change* (Cambridge: Cambridge University Press, 2011); Nicholas Chan, "Climate Contributions and the Paris Agreement: Fairness and

Equity in a Bottom-Up Architecture," *Ethics & International Affairs* 30, no. 3 (2016): 291–301.

60 Eric A. Posner and Cass R. Sunstein, "Should Greenhouse Gas Permits Be Allocated on a Per Capita Basis?" *California Law Review* 97 (2009): 51.

61 Alan Buis, "The Atmosphere: Getting a Handle on Carbon Dioxide," newsletter, NASA's Jet Propulsion Laboratory, 9 October 9, 2019, https://climate.nasa.gov/news/2915/the-atmosphere-getting-a-handle-on-carbon-dioxide/.

62 Eric A. Posner and Cass R. Sunstein, "Climate Change Justice," *Georgetown Law Journal* 96 (2008): 1567.

63 ECCC, *Global Greenhouse Gas Emissions: Canadian Environmental Sustainability Indicators* (Gatineau: ECCC, 2020), 5, fig. 1, https://www.canada.ca/content/dam/eccc/documents/pdf/cesindicators/global-ghg-emissions/2020/global-ghg-emissions-en.pdf.

64 Posner and Sunstein, "Greenhouse Gas Permits."

65 Government of Canada, *Canada's Green Plan: Canada's Green Plan for a Healthy Environment* (Ottawa: Supply and Services Canada, 1990), 99, https://cfs.nrcan.gc.ca/pubwarehouse/pdfs/24604.pdf.

66 See note 47 above.

67 Peter Zimonjic, "Trudeau Tells Trump Canada Is Disappointed by Withdrawal from Paris Climate Deal," CBC, 1 June 2017, https://www.cbc.ca/news/politics/trudeau-mckenna-trump-paris-deal-1.4142211.

68 ECCC, *Heathy Environment, Healthy Economy*, 31.

69 International Renewable Energy Agency, *Global Energy Transformation: A Roadmap to 2050* (Abu Dhabi: IRENA, 2019), https://www.irena.org/publications/2019/Apr/Global-energy-transformation-A-roadmap-to-2050-2019Edition.

70 Chris Sworder, Louisiana Salge, and Henri Van Soest, *The Global Cleantech Innovation Index 2017* (San Francisco: Cleantech Group and WWF, 2017), 31, http://info.cleantech.com/WWF-Index-2017_WWF-Index-2017-Submit.html.

71 Canadian Institute for Climate Choices, *Sink or Swim*, 12, 30.

72 ECCC, *Healthy Environment, Healthy Economy*, 36.

73 Canadian Institute for Climate Choices, *Sink or Swim*, 3.

74 Elinor Ostrom, "A Polycentric Approach for Coping with Climate Change," *Annals of Economics and Finance* 15, no. 1 (2014): 104, https://doi.org/10.1016/j.ecolecon.2019.02.005.

75 Ostrom, "Polycentric Approach," 105.
76 Stefano Carattini, Simon Levin, and Alessandro Tavoni, "Cooperation in the Climate Commons," *Review of Environmental Economics and Policy* 13, no. 2 (2019): 235.
77 Carattini, Levin, and Tavoni, "Cooperation," 232.
78 For example, see Nordhaus, "Climate Clubs," 1339. See also Barrett, *Why Cooperate?* And Jaccard, *Citizen's Guide.*
79 See, for example, Dale Beugin of the Ecofiscal Commission in House of Commons, *Clean Growth and Climate Change: How Canada Can Lead Internationally. Report of the Standing Committee on Environment and Sustainable Development* (Ottawa, April 2019), 40, https://www .ourcommons.ca/Content/Committee/421/ENVI/Reports/RP10403102 /envirp19/envirp19-e.pdf.
80 For a good overview of Innis, the staples thesis, and the empirical literature related to it, see Nancy Olewiler, "Canada's Dependence on Natural Capital Wealth: Was Innis Wrong?" *Canadian Journal of Economics* 50 No. 4 (2017): 927.
81 Olewiler, "Canada's Dependence," and Stephen Gordon, "Canada Should Specialize in Resource Development," *National Post*, 13 November 2017, https://nationalpost.com/opinion/stephen-gordon-canada -should-specialize-in-resource-development.
82 Olewiler, "Canada's Dependence," and Brendan Haley, "From Staples Trap to Carbon Trap: Canada's Peculiar Form of Carbon Lock-In," *Studies in Political Economy* 88 (2011): 97.
83 See, for example, Ann Carlos and Frank Lewis, "Property Rights, Competition and Depletion in the Eighteenth-Century Canadian Fur Trade: The Role of European Markets," *Canadian Journal of Economics* 32, no. 3 (1999): 705.
84 Brendan Haley, "The Staple Theory and the Carbon Trap," in *The Staple Theory @ 50: Reflections on the Lasting Significance of Mel Watkins "A Staple Theory of Economic Growth,"* ed. Jim Stanford (Canadian Centre for Policy Alternatives, 2014), 77.
85 Expert Panel on Sustainable Finance, *Final Report of the Expert Panel on Sustainable Finance: Mobilizing Finance for Sustainable Growth* (Gatineau: ECCC, 2019), https://publications.gc.ca/collections/collection_2019 /eccc/En4-350-2-2019-eng.pdf, 39, citing Peter Tertzakian, "Alberta: Why Scale and Policy Matter," *ARC Energy Research Institute*, 30 April 2019,

https://www.arcenergyinstitute.com/alberta-why-scale-and-policy
-matter/ (fig. 2).

86 Robson Fletcher, "Oil and Gas Surpasses Banking and Insurance as Share
of GDP in May," CBC News, 31 July 2018, https://www.cbc.ca/news
/anada/anada/oil-and-gas-gdp-growth-may-data-statscan-1.4768508.
For 20% of trade, see Government of Canada, "Market Snapshot: Crude
Oil – One of Canada's Top Exports Is Also One of the Most Globally
Traded Commodities," 12 June 2019, https://www.cer-rec.gc.ca/nrg
/ntgrtd/mrkt/snpsht/2019/06-02crdl-eng.html.

87 Natural Resources Canada, "Crude Oil Facts," https://www.nrcan
.gc.ca/crude-oil-facts/20064.

88 Natural Resources Canada, "Crude Oil Facts," (the largest is Venezuela,
followed by Saudi Arabia and then Canada). See also Olewiler,
"Canada's Dependence," 953.

89 Olewiler, "Canada's Dependence," 934.

90 Olewiler, "Canada's Dependence," 939.

91 Olewiler, "Canada's Dependence," 940–1.

92 Haley, "From Staples Trap," 120–1; Haley, "The Staple Theory."

93 Katrina Wyman, "From Fur to Fish: Reconsidering the Evolution of
Private Property," *NYU Law Review* 80 (2005): 117.

94 On the influence of powerful industries, particularly natural resource
industries, on Canadian environmental law, see David Boyd, *Unnatural
Law: Rethinking Canadian Environmental Law and Policy* (Vancouver:
UBC Press, 2003); Stepan Wood, Georgia Tanner, and Benjamin
Richardson, "What Ever Happened to Canadian Environmental
Law?" *Ecology Law Quarterly* 37 (2010): 981; Jason MacLean, "Striking
at the Root Problem of Canadian Environmental Law: Identifying
and Escaping Regulatory Capture," *Journal of Environmental Law and
Practice* 29 (2016): 111–28.

95 Jason MacLean, "Regulatory Capture and the Role of Academics in
Public Policymaking: Lessons from Canada's Environmental Regulatory
Review Process," *UBC Law Review* 52, no. 2 (June 2019): 479, 513.

96 Glen Hodgson, "'Peak Oil' Is Back – But Not on the Supply Side," *Globe
and Mail*, 23 August 2019, https://maritimesenergy.com/news/2019
-08-23/-peak-oil-is-back-but-not-on-the-supply-side-globe-mail. But see
Jaccard, *Citizen's Guide*, arguing that peak oil is not an important factor as
there are plenty of reserves left.

97 McKinsey & Company, *Global Energy Perspective 2019: Reference Case* (New York: McKinsey Solutions SPRL, 2019), https://www.mckinsey .com/~/media/McKinsey/Industries/Oil%20and%20Gas/Our%20 Insights/Global%20Energy%20Perspective%202019/McKinsey-Energy -Insights-Global-Energy-Perspective-2019_Reference-Case-Summary.ashx.

3 Discretion I: Picking the Wrong Tools

1 *References re Greenhouse Gas Pollution Pricing Act*, 2021 SCC 11.
2 Nicholas Stern, "Ethics, Equity and the Economics of Climate Change. Paper 2: Economics and Politics," *Economics and Philosophy* 30, no. 3 (2014): 489 (arguing that global agreements should be analysed on the basis of whether they are effective, efficient, and equitable). See also Canada's Ecofiscal Commission, *Supporting Carbon Pricing: How to Identify Policies that Genuinely Complement an Economy-Wide Carbon Price* (Montreal: Canada's Ecofiscal Commission, 2017), v, http://ecofiscal .ca/wp-content/uploads/2017/06/Ecofiscal-Commission-Report -Supporting-Carbon-Pricing-June-2017.pdf. Others use a range of criteria. Mark Jaccard, for example, writes about four criteria: economic efficiency, emissions effectiveness, political acceptability, and administrative feasibility. Mark Jaccard, Mikela Hein, and Tiffany Vass, *Is Win-Win Possible? Can Canada's Government Achieve Its Paris Commitment ... and Get Re-elected?* (Vancouver: School of Resource and Environmental Management, Simon Fraser University, 2016), 3, http://www.sfu.ca /content/dam/sfu/emrg/Publications/Research_Publications/2016%20 Jaccard,%20Hein%20&%20Vass%20-%20Is%20Win-Win%20Possible.pdf.
3 This broader notion of efficiency is related to the field of welfare economics, which examines how to assess the impact of broader policy shifts but also could involve non-welfarist approaches to assessing policy. For a discussion of economic and philosophical approaches to discussing policy and justice, see Amartya Sen, *The Idea of Justice* (Cambridge, MA: Belknap, 2009).
4 See, for example, William Nordhaus, "Climate Change: The Ultimate Challenge for Economics," *American Economic Review* 109, no. 6 (June 2019): 2003.
5 Martin L. Weitzman, "Can Negotiating a Uniform Carbon Price Help to Internalize the Global Warming Externality?" *Journal of the Association of Environmental and Resource Economists* 1, no. 1/2 (2014): 30.

6   See, for example, A. Denny Ellerman, Claudio Marcantonini, and
    Aleksandar Zaklan, "The European Union Emissions Trading System:
    Ten Years and Counting," *Review of Environmental Economics and Policy* 10,
    no. 1 (2016): 98.

7   Mark Jaccard, *The Citizen's Guide to Climate Success* (Cambridge:
    Cambridge University Press, 2020).

8   Canada's Ecofiscal Commission, *Bridging the Gap: Real Options for Meeting
    Canada's 2030 GHG Target* (Montreal: Canada's Ecofiscal Commission,
    2019), 16, https://ecofiscal.ca/wp-content/uploads/2019/11/Ecofiscal
    -Commission-Bridging-the-Gap-November-27–2019-FINAL.pdf. See also
    Jaccard, Hein, and Vass, *Is Win-Win Possible?*, 23 (putting the price at $200
    to $215); Parliamentary Budget Officer, *Closing the Gap: Carbon Pricing for
    the Paris Target* (Ottawa: PBO, 13 June 2019), https://www.pbo-dpb
    .gc.ca/web/default/files/Documents/Reports/2019/Paris_Target
    /Paris_Target_EN.pdf (estimates we need an added $52 in the carbon
    -price-added current policies).

9   Shawn McCarthy, "Canada's Oil Sands Producers Race to Cut Carbon
    Footprint," *Globe and Mail*, 23 August 2019, https://www.theglobeandmail
    .com/business/article-canadas-oil-sands-producers-race-to-cut-its
    -carbon-footprint/.

10 Nordhaus, "Climate Change," 2019.

11 Parliamentary Budget Officer, *Closing the Gap*, 11.

12 There is a long line of literature detailing the relative inefficiency of
    "command and control" environmental regulations. See, for example,
    Neil Gunningham, "Environment Law, Regulation and Governance:
    Shifting Architectures," *Journal of Environmental Law* 21, no. 2
    (2009): 182–4.

13 Jaccard, *Citizen's Guide*, 110–14.

14 Jaccard, Hein, and Vass, *Is Win-Win Possible?*.

15 Jaccard, Hein, and Vass, *Is Win-Win Possible?*, 23.

16 Nordhaus, "Climate Change," 2019; Seth Klein, *A Good War: Mobilizing
    Canada for the Climate Emergency* (Vancouver: ECW Press, 2020).

17 Andrew Leach, "What's Really Holding Back the Oilsands? It's Not the
    Bill of Goods You're Being Sold," CBC News, 4 September 2019, https://
    www.cbc.ca/news/anada/anada/road-ahead-oilsands-future-andrew
    -leach-1.5268556. See also Expert Panel on Sustainable Finance, *Final
    Report of the Expert Panel on Sustainable Finance: Mobilizing Finance for
    Sustainable Growth* (Gatineau: ECCC, 2019); and Shawn McCarthy,

"Canada's Oil Sands Producers Race to Cut Carbon Footprint," *Globe and Mail*, 23 August 2019.

18 For a review of studies of voluntary programs, see Jonathan C. Borck and Cary Coglianese, "Voluntary Environmental Programs: Assessing Their Effectiveness," *Annual Review of Environment and Resources* 34 (2009).

19 Ecofiscal Commission, *Bridging the Gap*; Jaccard, *Citizen's Guide*, 125; Nicholas Rivers and Randall Wigle, "Reducing Greenhouse Gas Emissions in Transport: All in One Basket?" *University of Calgary School of Public Policy Briefing Paper* 11, no. 5 (2018).

20 For good discussions of the early path of climate policy in Canada, see Jeffrey Simpson, Mark Jaccard, and Nic Rivers, *Hot Air: Meeting Canada's Climate Change Challenge* (Toronto: McClelland and Stewart, 2007), as well as Kathryn Harrison "The Road Not Taken: Climate Change Policy in Canada and the United States," *Global Environmental Politics* 7, no. 4 (November 2007): 92–117.

21 Harrison, "The Road Not Taken," 109–10. More generally, see Simpson, Jaccard, and Rivers, *Hot Air*; Heather A. Smith, "Political Parties and Canadian Climate Change Policy," in "Electoral Politics and Policy: Annual John W. Holmes Issue on Canadian Foreign Policy," ed. David Black, Brian Bow, and Pierre Martin, special issue, *International Journal* 64, no. 1 (2008/2009): 49.

22 Auditor General of Canada, *Report of the Commissioner of the Environment and Sustainable Development. Chapter 2: Meeting Canada's 2020 Climate Change Commitments* (Ottawa: Auditor General of Canada, 2012), 33, http://publications.gc.ca/collections/collection_2012/bvg-oag/FA1-2 -2012-1-2-eng.pdf.

23 Auditor General of Canada, *Report of the Commissioner of the Environment and Sustainable Development. Chapter 1: Mitigating Climate Change* (Ottawa: Auditor General of Canada, 2014), 6, http://publications.gc.ca /collections/collection_2014/bvg-oag/FA1-2-2014-1-1-eng.pdf.

24 Government of Canada, "Pan-Canadian Framework on Clean Growth and Climate Change Second Annual Report: Section 4," last modified 8 August 2019, https://www.canada.ca/en/environment-climate-change /services/climate-change/pan-canadian-framework-reports/second -annual-report/section-4.html.

25 ECCC, *Healthy Environment, Healthy Economy*.

26 Canada, *2030 Emissions Reduction Plan* (Ottawa: 2022).

27 For an overview of carbon pricing in Canada, see Tracy Snoddon
and Trevor Tombe, "Analysis of Carbon Tax Treatment in Canada's
Equalization Program," *Canadian Public Policy* 45, no. 3 (2019): 377–92,
https://www.muse.jhu.edu/article/734298.

28 Act to Increase the Number of Zero-Emission Motor Vehicles in Quebec
in order to Reduce Greenhouse Gas and Other Pollutant Emissions,
CQLR, c A-33.02; British Columbia, *Building a Cleaner, Stronger BC: 2019
Climate Change Accountability Report* (Province of British Columbia, 2020),
15, https://www2.gov.bc.ca/assets/gov/environment/climate-change
/action/progress-to-targets/2019-climatechange-accountability-report
-web.pdf. See also British Columbia, "Renewable & Low Carbon Fuel
Requirements Regulation," last modified 30 March 2020, https://www2
.gov.bc.ca/gov/content/industry/electricity-alternative-energy
/transportation-energies/renewable-low-carbon-fuels; Climate Change
and Emissions Management Act, SA 2003, c C-16.7, subsequently
renamed The Emissions Management and Climate Resilience Act, SA
2003, c E-7.8. See also Jaccard, *Citizen's Guide*, 110.

29 ECCC, *Estimated Results of the Federal Carbon Pollution Pricing System*
(Gatineau: ECCC, 2018), 1, https://www.canada.ca/content/dam
/eccc/documents/pdf/reports/estimated-impacts-federal-system
/federal-carbon-pollution-pricing-system_en.pdf. See also Mark Jaccard,
"Canadian Carbon Pricing Confusions," *Sustainability Suspicions* (blog),
30 April 2018, http://markjaccard.blogspot.com/2018/04/canadian
-carbon-pricing-confusions.html.

30 S. Langlois-Bertrand, K. Vaillancourt, L. Beaumier, M. Pied, O. Bahn, and
N. Mousseau, *Canadian Energy Outlook 2021 – Horizon 2060* (Montreal:
Institut de l'énergie Trottier and e3c Hub, 2021), 92.

31 Nordhaus, "Climate Change," 1994; Nicholas Stern, "Ethics, Equity and
the Economics of Climate Change Paper 1: Science and Philosophy,"
Economics and Philosophy 30, no. 3 (2014): 431, https://doi.org/10.1017
/S0266267114000297.

32 ECCC, "Technical Update to Environment and Climate Change Canada's
Social Cost of Greenhouse Gas Estimates," March 2016, https://ec.gc.ca
/cc/default.asp?lang=En&n=BE705779-1.

33 ECCC, "Annex: Pricing Carbon Pollution," in *Healthy Environment,
Healthy Economy*, 2.

34 This use of cost-benefit analysis raises other concerns such as
overestimation of the costs of adopting environmental regulations.

Sustainable Prosperity, *Overestimating the Costs of Compliance with Environmental Regulation*, November 2015, 19, https://institute .smartprosperity.ca/sites/default/files/publications/files /Overestimating.pdf.

35 For differing views of the approach to discount rates, see Nordhaus, "Climate Change," 2005; Nicholas Stern, "Ethics, Equity and the Economics of Climate Change Paper 2: Economics and Politics," *Economics and Philosophy* 30, no. 3 (2014): 446, 466.

36 Nordhaus, "Climate Change," 2019.

37 For estimates of the social cost of carbon at various discount rates, see Nordhaus, "Climate Change," table 2.

38 Martin L. Weitzman, "Fat Tails and the Social Cost of Carbon," *American Economic Review* 104, no. 5 (2014): 544 (calling this extreme effect the "dismal theorem").

39 Weitzman, "Fat Tails," 544.

40 Weitzman, "Fat Tails," 546.

41 Nordhaus, "Climate Change," 2000.

42 Nordhaus, "Climate Change," 1996.

43 Stern, "Paper 2," 451.

44 Stern, "Paper 2," 475.

45 British Columbia, "Province Releases Greenhouse Gas Numbers for 2017," 9 September 2019, https://news.gov.bc.ca/releases/2019ENV0097 -001737.

46 Pembina Institute, "LNG Legislation Complicates BC's Climate Challenge," news release, 26 March 2019, https://www.pembina.org /media-release/bc-lng-legislation.

47 Government of Canada, "Ontario and Pollution Pricing," last modified 7 March 2019, https://www.canada.ca/en/environment-climate -change/services/climate-change/pricing-pollution-how-it-will-work /ontario.html.

48 Government of Ontario, "A Made-in-Ontario Environment Plan," https://www.ontario.ca/page/made-in-ontario-environment-plan.

49 Dave Sawyer and Seton Stiebert, *The Cost Implications of Ontario's Environment Plan to Reduce Greenhouse Gas Emissions* (EnviroEconomics/ Canadians for Clean Prosperity, 2019), 1, https://www.enviroeconomics .org/single-post/2019/06/04/the-cost-implications-of-ontario-s -environment-plan-to-reduce-greenhouse-gas-emissions.

50 Trevor Tombe, "No, Alberta Didn't Scrap Its Carbon Tax," *Globe and Mail*, 28 May 2019, https://www.theglobeandmail.com/opinion/article -jason-kenney-is-not-really-the-carbon-tax-killer-he-claims-to-be/.

51 *References re Greenhouse Gas Pollution Pricing Act*, 2021 SCC 11.

52 David R. Boyd, *Unnatural Law: Rethinking Canadian Environmental Law and Policy* (Vancouver: UBC Press, 2003), 292.

53 Lynda Collins and Lorne Sossin, "In Search of an Ecological Approach to Constitutional Principles and Environmental Discretion in Canada," *UBC Law Review* 52, no. 1 (2019): 293. See also Andrew Green, "Delegation and Consultation: How the Administrative State Functions and the Importance of Rules," in *Administrative Law in Context*, 3rd ed, ed. Colleen M. Flood and Lorne Sossin (Toronto: Emond, 2018), 307–40; Jocelyn Stacey, "The Environmental Emergency and the Legality of Discretion in Environmental Law," *Osgoode Hall Law Journal* 52, no. 3 (2015): 985–1028; Nigel Bankes, Sharon Mascher, and Martin Olszynski, "Can Environmental Laws Fulfill Their Promise? Stories from Canada," *Sustainability* 6, no. 9 (2014): 6024–48.

54 Collins and Sossin, "Ecological Approach," 294.

55 Stacey, "Environmental Emergency."

56 Green, "Delegation and Consultation."

57 For a discussion of regulatory capture and Canadian environmental law, see, for example, Boyd, *Unnatural Law*; Jason MacLean, "Regulatory Capture and the Role of Academics in Public Policymaking: Lessons from Canada's Environmental Regulatory Review Process," *UBC Law Review* 52, no. 2 (2019): 479–552; Jason MacLean, "Striking at the Root Problem of Canadian Environmental Law: Identifying and Escaping Regulatory Capture," *Journal of Environmental Law and Practice* 29 (2016): 111–28. See also Stacey, "Environmental Emergency."

58 Boyd, *Unnatural Law*, 256.

59 MacLean, "Regulatory Capture," 498; Stepan Wood, Georgia Tanner, and Benjamin J. Richardson, "What Ever Happened to Canadian Environmental Law?" *Ecology Law Quarterly* 37, no. 4 (2010): 1013.

60 Canadian Institute for Climate Choices, *Tip of the Iceberg: Navigating the Known and Unknown Costs of Climate Change for Canada* (Ottawa: CICC, 2020), https://climatechoices.ca/wp-content/uploads/2020/12/Tip-of -the-Iceberg-_-CoCC_-Institute_-Full.pdf.

61 Nordhaus, "Climate Change."

62 Scott Barrett, "Solar Geoengineering's Brave New World: Thoughts on the Governance of an Unprecedented Technology," *Review of Environmental Economics and Policy* 8, no. 2 (2014): 249–69.

4 Discretion II: Helping Everyone Helps No One

 1 Kathryn Harrison, *Passing the Buck: Federalism and Canadian Environmental Policy* (Vancouver: UBC Press, 2000).
 2 Roger Pielke, Jr., *The Climate Fix: What Scientists and Politicians Won't Tell You about Global Warming* (New York: Basic Books, 2010). See also Michael J. Trebilcock, *Dealing with Losers: The Political Economy of Policy Transitions* (Oxford: Oxford University Press, 2014), 128.
 3 Natural Resources Canada, *Energy Fact Book: 2019–2020* (Ottawa: Natural Resources Canada, 2020), 2, 6, 13, https://www.nrcan.gc.ca/sites/www .nrcan.gc.ca/files/energy/energy-factbook_EN-feb14-2020.pdf.
 4 Government of Canada, foreword to *Pan-Canadian Framework on Clean Growth and Climate Change: Canada's Plan to Address Climate Change and Grow the Economy* (Gatineau: ECCC, 2016), http://publications .gc.ca/collections/collection_2017/eccc/En4-294-2016-eng.pdf. See also Smart Prosperity Institute, *New Thinking: Canada's Roadmap to Smart Prosperity* (Ottawa: Smart Prosperity Institute, 2016), 2, https://institute .smartprosperity.ca/sites/default/files/newthinking.pdf (discussing the possibility of a "win-win" path, 2).
 5 For a good explanation of the costs and benefits of using CGE models, see Jared C. Carbone and Nicholas Rivers, "The Impacts of Unilateral Climate Policy on Competitiveness: Evidence from Computable General Equilibrium Models," *Review of Environmental Economics and Policy* 11, no. 1 (2017): 24–42.
 6 Elizabeth Beale et al., *Provincial Carbon Pricing and Competitiveness Pressures: Guidelines for Business and Policymakers* (Montreal: Canada's Ecofiscal Commission, 2015), 3, https://ecofiscal.ca/wp-content /uploads/2015/11/Ecofiscal-Commission-Carbon-Pricing -Competitiveness-Report-November-2015.pdf.
 7 Chris Bataille, Benjamin Dachis, and Nic Rivers, *Pricing Greenhouse Gas Emissions: The Impact on Canada's Competitiveness*, commentary no. 280 (Toronto: C.D. Howe Institute, 2009), 8, https://www.cdhowe.org/sites

/default/files/attachments/research_papers/mixed//commentary_280
.pdf; Nic Rivers, "Impacts of Climate Policy on the Competitiveness of
Canadian Industry: How Big and How to Mitigate?" *Energy Economics* 32,
no. 5 (2010): 1096–7.

8 Parliamentary Budget Officer, *Canada's Greenhouse Gas Emissions:
Developments, Prospects and Reductions* (Ottawa: PBO, 2016), 25, https://
www.pbo-dpb.gc.ca/web/default/files/Documents/Reports/2016
/ClimateChange/PBO_Climate_Change_EN.pdf; Nick Macaluso
et al., "The Impact of Carbon Taxation and Revenue Recycling on U.S.
Industries," *Climate Change Economics* 9, no. 1 (2018); Toon Vandyck
et al., "A Global Stocktake of the Paris Pledges: Implications for
Energy Systems and Economy," *Global Environmental Change* 41 (2016):
46–63; Andries F. Hof et al., "Global and Regional Abatement Costs of
Nationally Determined Contributions (NDCs) and of Enhanced Action
to Levels Well Below 2°C and 1.5°C," *Environmental Science & Policy*
71 (2017): 30–40; Parliamentary Budget Officer, *Closing the Gap: Carbon
Pricing for the Paris Target* (Ottawa: PBO, 2019), 15, https://www.pbo
-dpb.gc.ca/web/default/files/Documents/Reports/2019/Paris_Target
/Paris_Target_EN.pdf.

9 Government of Canada, "Fall 2018 Update: Estimated Impacts of the
Federal Pollution Pricing System," last modified 31 January 2019,
https://www.canada.ca/en/environment-climate-change/services
/climate-change/pricing-pollution-how-it-will-work/fall-2018-update
-estimated-impacts-federal-pollution-pricing-system.html.

10 ECCC, "Annex: Modelling and Analysis," in *Healthy Environment, Healthy
Economy.*

11 Akio Yamazaki, "Jobs and Climate Policy: Evidence from British
Columbia's Revenue-Neutral Carbon Tax," *Journal of Environmental
Economics and Management* 83 (2017): 197–216; Deven Azevedo, Hendrik
Wolff, and Akio Yamazaki, "Do Carbon Taxes Kill Jobs? Firm-Level
Evidence from British Columbia," Clean Economy Working Paper series
WP 18–08, Smart Prosperity Institute, University of Ottawa, Ottawa,
March 2019, https://institute.smartprosperity.ca/sites/default/files
/docarbontaxeskilljobsmarch2019.pdf; Chi Man Yip, "On the Labor
Market Consequences of Environmental Taxes," *Journal of Environmental
Economics and Management* 89 (2018): 136–52.

12 Carbone and Rivers, "Impacts of Unilateral Climate Policy," 24–5.

13 Parliamentary Budget Officer, *Developments, Prospects and Reductions*, 25–6.

14 In a survey of over fifty CGE models, Carbone and Rivers found only a small effect on welfare of a unilateral carbon tax, between 0 and 2% for a 20% reduction in emissions. Carbone and Rivers, "Impacts of Unilateral Climate Policy," 35.

15 See, for example, Robin Edger et al., *Healthy Recovery Plan: For a Safe and Sustainable Future* (Toronto: Canadian Association of Physicians for the Environment, 2020), 5, https://cape.ca/wp-content/uploads/2020/07/CAPE_Report_EN_HealthyRecoveryPlan.pdf.

16 Nicholas Stern, *The Economics of Climate Change: The Stern Review* (Cambridge: Cambridge University Press, 2007), xv.

17 Nicholas Stern, "Ethics, Equity and the Economics of Climate Change Paper 1: Science and Philosophy," *Economics and Philosophy* 30, no. 3 (2014): 436.

18 Canadian Institute for Climate Choices, *Tip of the Iceberg: Navigating the Known and Unknown Costs of Climate Change for Canada* (Ottawa: CICC, 2020), iii, https://climatechoices.ca/wp-content/uploads/2020/12/Tip-of-the-Iceberg-_-CoCC_-Institute_-Full.pdf.

19 F. Warren and N. Lulham, eds., *Canada in a Changing Climate: National Issues Report* (Ottawa: Government of Canada, 2021), 349, https://changingclimate.ca/site/assets/uploads/sites/3/2021/05/National-Issues-Report_Final_EN.pdf.

20 Bataille, Dachis, and Rivers, *Pricing Greenhouse Gas Emissions*, 2.

21 Mohammad S. Masnadi et al., "Global Carbon Intensity of Crude Oil Production," *Science* 361, no. 6405 (2018): 851.

22 Cenovus Energy, "Open Letter to Canadians," accessed 29 July 2020, https://www.cenovus.com/news/our-stories/open-letter-to-canadians.html. See also "The Environment Is Canada's Biggest Wedge Issue," *Economist*, 25 July 2019, https://www.economist.com/special-report/2019/07/25/the-environment-is-canadas-biggest-wedge-issue.

23 Natural Resources Canada, "Industrial Energy Intensity by Industry," accessed 20 July 2020, https://oee.nrcan.gc.ca/corporate/statistics/neud/dpa/showTable.cfm?type=HB&sector=agg&juris=00&rn=6&page=0.

24 Canada's Ecofiscal Commission, *Choose Wisely: Options and Trade-Offs in Recycling Carbon Pricing Revenues* (Montreal: Canada's Ecofiscal

Commission, 2016), 8, http://ecofiscal.ca/wp-content/uploads/2016/04
/Ecofiscal-Commission-Choose-Wisely-Carbon-Pricing-Revenue
-Recycling-Report-April-2016.pdf.

25 World Bank Group, "Trade (% of GDP) – Canada, Japan, United States,"
World Bank Open Data, accessed 20 July 2020, https://data.worldbank
.org/indicator/NE.TRD.GNFS.ZS?end=2018&locations=CA-JP-US
&start=2018&view=bar.

26 Meredith L. Fowlie, Mar Reguant, and Stephen P. Ryan, "Measuring
Leakage Risk," report, Public Workshop on Emissions Leakage Studies
for Cap-and-Trade Program, California Air Resources Board, Sacramento,
CA, 18 May 2016), 6, https://ww2.arb.ca.gov/sites/default/files
/classic//cc/capandtrade/meetings/20160518/ucb-intl-leakage.pdf.

27 Beale et al., *Competitiveness Pressures*, 13–5. Different measures of trade
exposure may be used: Bataille, Dachis, and Rivers, *Pricing Greenhouse
Gas Emissions*, 3–4.

28 The Ecofiscal Commission was unable to use actual data because of
confidentiality and instead used data related to a CGE model. Beale et al.,
Competitiveness Pressures, 4.

29 Beale et al., *Competitiveness Pressures*, 4. See also Carlos A. Murillo,
*Tipping the Scales: Assessing Carbon Competitiveness and Leakage Potential for
Canada's EITEIs* (Ottawa: Conference Board of Canada, 2019), 18, https://
www.conferenceboard.ca/temp/2593155d-38ed-4e50-9384-41663d2ebe7e
/10485_TippingtheScales-RPT.pdf.

30 Carbone and Rivers, "Impacts of Unilateral Climate Policy," 25.

31 Antoine Dechezleprêtre and Misato Sato, "The Impacts of Environmental
Regulations on Competitiveness," *Review of Environmental Economics and
Policy* 11, no. 2 (2017): 183–206; Misato Sato and Antoine Dechezleprêtre,
"Asymmetric Industrial Energy Prices and International Trade," *Energy
Economics* 52, no. S1 (2015): S130–S141.

32 See, for example, Bataille, Dachis, and Rivers, *Pricing Greenhouse Gas
Emissions*, 5–9; Rivers, "Impacts of Climate Policy," 1096–8; National
Round Table on the Environment and the Economy, *Achieving 2050:
A Carbon Pricing Policy for Canada* (Ottawa: National Round Table on the
Environment and the Economy, 2009), https://publications.gc.ca
/collections/collection_2009/trnee-nrtee/En134-43-1-2009E.pdf, 76.

33 Joseph E. Aldy and William A. Pizer, "The Competitiveness Impacts
of Climate Change Mitigation Policies," *Journal of the Association of*

Environmental and Resource Economists 2, no. 4 (2015): 566; Fowlie,
Reguant, and Ryan, "Measuring Leakage Risk," 7.

34 Sato and Dechezleprêtre, "Asymmetric Industrial Energy Prices," S139.
35 Carbone and Rivers, "Impacts of Unilateral Climate Policy," 33. See also
Macaluso et al., "Impact of Carbon Taxation," 18–21, 29.
36 Rivers, "Impacts of Climate Policy," 1097.
37 See Rivers, "Impacts of Climate Policy," 1093, and Carbone and Rivers,
"Impacts of Unilateral Climate Policy," 31, discussing prior studies.
See also Murillo, *Tipping the Scales*, 18; Chris Ragan and Jason Dion,
"Problematic New Study Overestimates Effects of Carbon Pricing in
Canada," Canada's Ecofiscal Commission, 5 November 2019, https://
ecofiscal.ca/2019/11/05/problematic-new-study-overestimates-effects
-carbon-pricing-canada/; Ross McKitrick, Elmira Aliakbari, and Ashley
Stedman, "The Impact of the Federal Carbon Tax on the Competitiveness
of Canadian Industries," Fraser Institute, 22 August 2019, https://www
.fraserinstitute.org/studies/impact-of-the-federal-carbon-tax-on-the
-competitiveness-of-canadian-industries.
38 Meredith Fowlie, Mar Reguant, and Stephen P. Ryan, "Market-Based
Emissions Regulation and Industry Dynamics," *Journal of Political
Economy* 124, no. 1 (2016): 253, 299–300. See also Carbone and Rivers,
"Impacts of Unilateral Climate Policy," 36.
39 Trebilcock, *Dealing with Losers*, 1.
40 See generally Louis Kaplow, "Transition Policy: A Conceptual
Framework," *Journal of Contemporary Legal Issues* 13, no. 1 (2003): 161–210;
Trebilcock, *Dealing with Losers*, 131–5.
41 Note that these companies still face an incentive to reduce emissions
because they could reduce emissions and sell these permits. However,
they do not pay for their current level of emissions.
42 See the Regulatory Impact Analysis Statement regarding the Regulations
Respecting Reduction in the Release of Methane and Certain Volatile
Organic Compounds (Upstream Oil and Gas Sector), (2017) C. Gaz. I,
2075, 2082, http://www.gazette.gc.ca/rp-pr/p1/2017/2017-05-27
/pdf/g1-15121.pdf, and Margo McDiarmid, "Federal Government Seeks
to Push Back Methane Reduction Regulations by Up to 3 Years," CBC
News, 20 April 2017, https://www.cbc.ca/news/politics/methane
-emissions-regulations-changes-1.4078468.
43 Trebilcock, *Dealing with Losers*, 131–5. See also Jonathan Remy Nash and
Richard L. Revesz, "Grandfathering and Environmental Regulation: The

Law and Economics of New Source Review," *Northwestern University Law Review* 101, no. 4 (2007): 1724–33.

44 Dale Marshall, "Reducing Canada's Methane Emissions Should Be a No-Brainer," Environmental Defence, 26 July 2017, https://environmentaldefence.ca/2017/07/26/reducing-canadas-methane-emissions-no-brainer.

45 Nash and Revesz, "Grandfathering and Environmental Regulation," 1708–10. See also Bruce R. Huber, "Transition Policy in Environmental Law," *Harvard Environmental Law Review* 35, no. 1 (2011): 93–4; Jonathan Remy Nash, "Allocation and Uncertainty: Strategic Responses to Environmental Grandfathering," *Ecology Law Quarterly* 36, no. 4 (2009): 835.

46 Kaplow, "Transition Policy," 186; but see Christopher T. Wonnell, "The Noncompensation Thesis and Its Critics: A Review of This Symposium's Challenges to the Argument for Not Compensating Victims of Legal Transitions," *Journal of Contemporary Legal Issues* 13, no. 1 (2003): 295 (noting that the rationality assumption may not work for individuals as opposed to firms).

47 Maria Damon, Daniel H. Cole, Elinor Ostrom, and Thomas Sterner, "Grandfathering: Environmental Uses and Impacts" *Review of Environmental Economics and Policy* 13, no. 1 (2019): 38.

48 Kaplow, "Transition Policy," 198. See also Trebilcock, *Dealing with Losers*, 19 (noting that, on the public choice view, if transition policy is politically desirable, "low-visibility, off-budget strategies such as grandfathering or phased or delayed implementation are likely to be preferred, and are generally likely to be biased in favor of concentrated and politically well-organized interests" while there is less likely political support for "the compensation or mitigation of isolated, widely-dispersed, or temporally attenuated losses," 19); Nash and Revesz, "Grandfathering and Environmental Regulation," 1725.

49 Huber, "Transition Policy," 110 (noting that these concentrated interests and environmental groups may agree on the resulting payments in the form of a Baptist and Bootlegger coalition).

50 Sarah Dobson and Jennifer Winter, "Assessing Policy Support for Emissions-Intensive and Trade-Exposed Industries," *School of Public Policy Publications* 11, no. 28 (2018): 39.

51 Fowlie, Reguant, and Ryan, "Measuring Leakage Risk," 5.

52 Beale et al., *Competitiveness Pressures*, 17; National Round Table on the Environment and the Economy, *Achieving 2050*, 75–6.

53 ECCC, *Canada's Fourth Biennial Report on Climate Change* (Gatineau: ECCC, 2019), 1, https://unfccc.int/sites/default/files/resource/br4_final_en.pdf.
54 Kathryn Harrison, "A Tale of Two Taxes: The Fate of Environmental Tax Reform in Canada," *Review of Policy Research* 29, no. 3 (2012): 387; Kathryn Harrison "The Road Not Taken: Climate Change Policy in Canada and the United States," *Global Environmental Politics* 7, no. 4 (2007): 107.
55 Marc Lee, *A Critical Look at BC's New Tax Breaks and Subsidies for LNG* (Vancouver: Canadian Centre for Policy Alternatives, 2019), 2, https://www.policyalternatives.ca/sites/default/files/uploads/publications/BC%20Office/2019/05/CCPA_BC%20Critiquing%20the%20LNG%20Canada%20agreement_FINAL_190506.pdf.
56 Nicholas Rivers and Brandon Schaufele, "The Effect of Carbon Taxes on Agricultural Trade," *Canadian Journal of Agricultural Economics* 63, no. 2 (2015): 235, 236.
57 Environmental Commissioner of Ontario, *Introduction to Cap and Trade in Ontario: Appendix A to the ECO's Greenhouse Gas Progress Report 2016* (Toronto: ECO, 2016), 10, https://www.auditor.on.ca/en/content/reporttopics/envother/env16_other/Appendix-A-Introduction-to-Cap-and-Trade-in-Ontario.pdf.
58 Government of Ontario, "Archived – Cap and Trade: Program Overview," 2 June 2016, https://www.ontario.ca/page/cap-and-trade-program-overview; Government of Ontario, *Methodology for the Distribution of Ontario Emission Allowances Free of Charge*, December 2016, 6, https://files.ontario.ca/methodology_for_the_distribution_of_ontario_emission_allowances_free_of_charge.pdf.
59 Beale et al., *Competitiveness Pressures*, 14.
60 Dobson and Winter, "Assessing Policy Support," 14.
61 Government of Canada, "Update on the Output-Based Pricing System: Technical Backgrounder," last modified 27 July 2018, https://www.canada.ca/en/services/environment/weather/climatechange/climate-action/pricing-carbon-pollution/output-based-pricing-system-technical-backgrounder.html.
62 Dobson and Winter, "Assessing Policy Support," 39.
63 Mark Jaccard, Mikela Hein, and Tiffany Vass, *Is Win-Win Possible? Can Canada's Government Achieve Its Paris Commitment … and Get Re-elected?* (Vancouver: School of Resource and Environmental Management, Simon

Fraser University, 2016), 5, http://www.sfu.ca/content/dam/sfu/emrg/Publications/Research_Publications/2016%20Jaccard,%20Hein%20&%20Vass%20-%20Is%20Win-Win%20Possible.pdf.

64 See Dobson and Winter, "Assessing Policy Support," 5–30.

65 Section 31(8) of the Climate Change Mitigation and Low-Carbon Economy Act, 2016, S.O. 2016, c. 7 (Can.), states: "Before January 1, 2021, the Minister shall make available to the public an outline that describes how the distribution of Ontario emission allowances free of charge will be phased out as Ontario makes the transition to a low-carbon economy."

66 Carbone and Rivers, "Impacts of Unilateral Climate Policy," 28; Christoph Böhringer, Jared C. Carbone, and Thomas F. Rutherford, "Unilateral Climate Policy Design: Efficiency and Equity Implications of Alternative Instruments to Reduce Carbon Leakage," *Energy Economics* 34, no. S2 (2012): S216.

67 Epps and Green, *Reconciling Trade*. See also Fowlie, Reguant, and Ryan, "Market-Based Emissions," 287.

68 See generally Tracey Epps and Andrew Green, *Reconciling Trade and Climate: How the WTO Can Help Address Climate Change* (Cheltenham: Edward Elgar, 2010); Paul-Erik Veel, "Carbon Tariffs and the WTO: An Evaluation of Feasible Policies," *Journal of International Economic Law* 12, no. 3 (2009): 749–800; Jaccard, *Citizen's Guide*, 71–5. In fact, this prohibition on protectionism under international trade rules may help address a different form of rent-seeking by industry where they seek to use environmental rules for economic as opposed to environmental gain.

69 Rivers, "Impacts of Climate Policy," 1101.

70 William Nordhaus, "Climate Clubs: Overcoming Free-Riding in International Climate Policy," *American Economic Review* 105, no. 4 (2015): 1339–70. See also Jaccard, *Citizen's Guide*, 22, 71–5.

71 Fowlie, Reguant, and Ryan, "Measuring Leakage Risk," 5; Rivers and Schaufele, "Agricultural Trade," 253–4.

72 Rivers, "Impacts of Climate Policy," 1103; Fowlie, Reguant, and Ryan, "Measuring Leakage Risk," 5; Böhringer, Carbone, and Rutherford, "Unilateral Climate Policy Design," S211–S212; Rivers and Schaufele, "Agricultural Trade," 253–4.

73 Carolyn Fischer and Alan K. Fox, "Comparing Policies to Combat Emissions Leakage: Border Carbon Adjustments Versus Rebates," *Journal of Environmental Economics and Management* 64, no. 2 (2012): 214.

74 Böhringer, Carbone, and Rutherford, "Unilateral Climate Policy Design," S216 (noting that beyond the BTA, an output-based system may be the best but "at the same time, the efficiency gains from output-based allocation are rather limited and, as a result, may not be worth the trouble to design and implement it in practice," S216).

75 Macaluso et al., "Impact of Carbon Taxation," 21; Rivers, "Impacts of Climate Policy," 1098–9.

76 Huber, "Transition Policy," 93. On exemptions generally, see Carbone and Rivers, "Impacts of Unilateral Climate Policy," 28. See also Rivers, "Impacts of Climate Policy," 1099–100.

77 Rivers and Schaufele, "Agricultural Trade," 253; Böhringer, Carbone, and Rutherford, "Unilateral Climate Policy Design," S212.

78 Government of Canada, "Clean Fuel Standard," last modified 24 April 2020, https://www.canada.ca/en/environment-climate-change/services/managing-pollution/energy-production/fuel-regulations/clean-fuel-standard.html.

79 Rivers, "Impacts of Climate Policy," 1099–100.

80 Rivers and Schaufele, "Agricultural Trade," 253; Böhringer, Carbone, and Rutherford, "Unilateral Climate Policy Design, S208, S213.

81 Hadrian Mertins-Kirkwood, "Heating Up, Backing Down: Evaluating Recent Climate Policy Progress in Canada," Working Paper no. 203, Canadian Centre for Policy Alternatives, Toronto, June 2019, 6, https://www.policyalternatives.ca/sites/default/files/uploads/publications/National%20Office/2019/06/Heating%20Up%2C%20Backing%20Down.pdf.

82 ECCC, *Healthy Environment, Healthy Economy*, 36.

83 ECCC, *Healthy Environment, Healthy Economy*, 30.

5 Diffusion: When Everyone's Responsible, No One's Responsible

1 Auditor General of Canada, *Perspectives on Climate Change Action in Canada – A Collaborative Report from Auditors General* (Ottawa: Auditor General of Canada, 2018), https://www.oag-bvg.gc.ca/internet/English/parl_otp_201803_e_42883.html.

2 Katherine Harrison, "The Fleeting Canadian Harmony on Carbon Pricing," *Policy Options*, 8 July 2019, https://policyoptions.irpp.org/magazines/july-2019/the-fleeting-canadian-harmony-on-carbon-pricing/.

3 See, for example, a discussion of the "west-east" split in Douglas Macdonald, *Carbon Province, Hydro Province: The Challenge of Canadian Energy and Climate Federalsim* (Toronto: University of Toronto Press, 2020), 6.

4 ECCC, *National Inventory Report 1990–2018: Greenhouse Gas Sources and Sinks in Canada* (Gatineau: ECCC, 2020), 11, table ES-4.

5 SaskWind, *Oil and Gas Production by Canadian Province. 2014* (2020), https://www.saskwind.ca/oil-gas-by-province.

6 Government of Canada, "Natural Gas Facts," last modified 2 July 2020, https://www.nrcan.gc.ca/science-data/data-analysis/energy-data -analysis/energy-facts/natural-gas-facts/20067.

7 Parliamentary Budgetary Officer, *Canada's Greenhouse Gas Emissions: Developments, Prospects and Reductions* (Ottawa: PBO, 2016), 15, https:// www.pbo-dpb.gc.ca/en/blog/news/Climate_Change_2016.

8 Nancy Olewiler, "Canada's Dependence on Natural Capital Wealth: Was Innis Wrong?" *Canadian Journal of Economics* 50, no. 4 (2017): 947.

9 Government of Alberta, "Unemployment Rate," 10 July 2020, https:// economicdashboard.alberta.ca/Unemployment#:~:text=PUBLISHED %20%2D%20Jul%2010%2C%202020,the%20same%20period%20in %202019; "The Environment Is Canada's Biggest Wedge Issue," *The Economist*, July 25, 2019, https://www.economist.com/special-report /2019/07/25/the-environment-is-canadas-biggest-wedge-issue.
   Elizabeth Beale et al., *Provincial Carbon Pricing and Competitiveness Pressures: Guidelines for Business and Policymakers* (Montreal: Canada's Ecofiscal Commission, 2015), 15, table 2, https://ecofiscal.ca/wp -content/uploads/2015/11/Ecofiscal-Commission-Carbon-Pricing -Competitiveness-Report-November-2015.pdf.

10 Government of Alberta, "Unemployment Rate," January 2021, https:// economicdashboard.alberta.ca/Unemployment#:~:text=PUBLISHED%20 %2D%20Jan%208%2C%202021,the%20same%20period%20in%202019; "Here's a Quick Glance at Unemployment Rates for August, by Province," *Toronto Star*, 10 September 2021, https://www.thestar.com /business/2021/09/10/heres-a-quick-glance-at-unemployment-rates -for-august-by-province.html.

11 Beale et al., *Provincial Carbon Pricing*, 15, table 2.

12 Canadian Institute for Climate Choices, *Sink or Swim: Transforming Canada's Economy for a Global Low-Carbon Future* (Ottawa: CICC, 2021), 59, https://climatechoices.ca/wp-content/uploads/2021/10/CICC-Sink-or -Swim-English-Final-High-Res.pdf.

13 Benjamin Israel, Jan Gorski, and Morrigan Simpson-Marran, "The Oilsands in a Decarbonizing Canada," Pembina Institute, October 2018, https://www.pembina.org/reports/oilsands-decarbonization -factsheet.pdf.

14 ECCC, *Canada's Fourth Annual Biennial Report on Climate Change* (Gatineau: ECCC, 2019), table 5.8, https://unfccc.int/sites/default/files /resource/br4_final_en.pdf.

15 ECCC, *Canada's Fourth Biennial Report*, 11.

16 Nichole Dusyk et al., *All Hands on Deck: An Assessment of Provincial, Territorial And Federal Readiness to Deliver a Safe Climate* (Pembina Institute, July 2021), 12.

17 Dusuk et al., *All Hands on Deck*, 4.

18 See Nicholas Stern, "Ethics, Equity and the Economics of Climate Change Paper 1: Science and Philosophy," *Economics and Philosophy* 30, no. 3 (2014): 397–444 (discussing the various philosophical approaches to addressing climate change).

19 Douglas Macdonald, Jochen Monstadt, and Kristine Kern, *Allocating Canadian Greenhouse Gas Emission Reductions amongst Sources and Provinces Learning from the European Union, Australia and Germany* (Toronto: University of Toronto, 2013), 53, https://environment.utoronto.ca /wp-content/uploads/2019/02/AllocatingGHGRe ductions2013.pdf.

20 Paul Boothe and Félix-A. Boudreault, *Sharing the Burden: Canadian GHG Emissions* (London: Lawrence National Centre for Policy and Management, 2016), https://www.ivey.uwo.ca/cmsmedia/2169603 /ghg-emissions-report-sharing-the-burden.pdf; Christoph Böhringer et al., "Sharing the Burden for Climate Change Mitigation in the Canadian Federation," *Canadian Journal of Economics / Revue Canadienne D'Economique* 48, no. 4 (2015): 1350–80, https://doi.org/10.1111 /caje.12152.

21 Böhringer et al., "Climate Change Mitigation," 1378.

22 See Boothe and Boudreault, *Sharing the Burden*; Eric A. Posner and Cass R. Sunstein, "Should Greenhouse Gas Permits Be Allocated on a Per Capita Basis?" *California Law Review* 97 (2009): 51–93.

23 See Böhringer et al., "Climate Change Mitigation," 1350–80.

24 Stern, "Ethics, Equity and Economics"; Böhringer et al., "Climate Change Mitigation," 1359, 1370. Philosopher John Rawls set out one of the most influential modern theories of justice in *A Theory of Justice* (Cambridge:

Belknap, 1971). He argued that individuals who had no information about who they were or what their position in society would be would adopt a system that would maximize the position of the worst-off in society.

25 Böhringer et al., "Climate Change Mitigation," 1357.
26 Böhringer et al., "Climate Change Mitigation," 1369, table 7.
27 Niels Anger et al., "Public Interest versus Interest Groups: A Political Economy Analysis of Allowance Allocation under the EU Emissions Trading Scheme," *International Environmental Agreements: Politics, Law and Economics* 16, no. 5 (2016) 621–38, https://doi.org/10.1007/s10784 -015-9285-6.
28 See Macdonald, Monstadt, and Kern, *Allocating Canadian Greenhouse Gas*, 51 (for a detailed discussion of attempts to reach agreement on burden sharing and the subsequent lack of explicit discussion of the issue).
29 ECCC, *Pan-Canadian Framework on Clean Growth and Climate Change: Canada's Plan to Address Climate Change and Grow the Economy* (Gatineau: ECCC, 2016), 5, http://publications.gc.ca/collections/collection_2017 /eccc/En4-294-2016-eng.pdf.
30 ECCC, "Pan-Canadian Framework," 3.
31 ECCC, "Pan-Canadian Framework," 8, 10.
32 ECCC, *Healthy Environment, Healthy Economy*, 84.
33 The federal government is in the process of reviewing its approach to ensuring provincial carbon pricing plans are equivalent following criticism that the approach has not been fair or consistent: Aaron Wherry, "Ottawa Submits New Greenhouse Gas Targets to UN, Plans Changes to Carbon-Pricing 'Benchmarks,'" *Globe and Mail*, 12 July 2021, https:// www.cbc.ca/news/politics/climate-change-carbon-emissions-united -nations-wilkinson-1.6097255.
34 Auditor General of Canada, *Perspectives on Climate Change*; ECCC, *National Inventory Report 1990–2017: Greenhouse Gas Sources and Sinks in Canada: Part 1*, 12, table ES-4 and fig. ES-8, https://publications.gc.ca /collections/collection_2019/eccc/En81-4-1-2017-eng.pdf; targets from ECCC, *Canada's Fourth Biennial Report*, 148, table A2.40.
35 Dusyk et al., *All Hands on Deck*, 4. See also ECCC, *Canada's Fourth Biennial Report*, 148, table A2.40.
36 Aaron Wherry, "By Claiming Ontario's Done Its 'Fair Share,' Doug Ford Pushes the Climate Burden West," *CBC News*, 25 April 2019, https://

www.cbc.ca/news/politics/doug-ford-climate-carbon-tax-emissions
-1.5108852.

37 Barry Saxifrage, "Surprise! Most of Canada Is On Track to Hit Our
2020 Climate Target," *National Observer*, 27 May 2019, https://www
.nationalobserver.com/2019/05/27/analysis/surprise-most-canada-track
-hit-our-2020-climate-target.

38 Macdonald, *Business and Environmental Politics.*

39 See discussion in chapter 10. See also Canadian Institute for Climate
Choices, *Charting Our Course: Bringing Clarity to Canada's Climate Policy
Choices on the Journey to 2050* (Ottawa: ECCC, 2020), 46.

40 See Cass Sunstein, "Incompletely Theorized Agreements," *Harvard Law
Review* 108, no. 7 (1998): 1733–72.

41 See MacDonald, Monstadt, and Kern, *Allocating Canadian Greenhouse Gas*,
34, discussing alternatives to an agreement on burden sharing.

42 MacDonald, Monstadt, and Kern, *Allocating Canadian Greenhouse Gas*, 20.

43 Macdonald, *Business and Environmental Politics in Canada*. See also
Macdonald, Monstadt, and Kern, *Allocating Canadian Greenhouse Gas*, 53.

44 Parliamentary Budget Officer, *Canada's Greenhouse Gas Emissions*, 28.

45 Macdonald, Monstadt and Kern, *Allocating Canadian Greenhouse Gas*, 34.

46 Posner and Sunstein, "Greenhouse Gas Permits," 86.

47 Boothe and Boudreault, *Sharing the Burden*, 6, table 1.

48 Boothe and Boudreault, *Sharing the Burden*, 5; Macdonald, Monstadt, and
Kern, *Allocating Canadian Greenhouse Gas.*

49 Macdonald, Monstadt, and Kern, *Allocating Canadian Greenhouse
Gas*, 78–9.

50 David Parkinson, "A Note to Western Canada: The Rest of the Country
Understands Tough Economic Times," *Globe and Mail*, 8 November 2019,
https://www.theglobeandmail.com/business/commentary/article-a
-note-to-brad-wall-the-rest-of-canada-understands-tough-economic/.

51 See Macdonald, Monstadt, and Kern, *Allocating Canadian Greenhouse Gas*,
132. They argue that the federal government's failure to use its fiscal
capacity to assist provinces or others has been a major barrier in the past
to getting a national agreement on allocating the costs of climate policy.

52 See Tracy R. Snoddon and Trevor Tombe, "Analysis of Carbon Tax
Treatment within Canada's Equalization Program," *Canadian Public Policy*
45, no. 3 (2019): 377–92.

53 Expert Panel on Sustainable Finance, *Final Report of the Expert Panel on
Sustainable Finance: Mobilizing Finance for Sustainable Growth* (Gatineau:

ECCC, 2019), http://publications.gc.ca/collections/collection_2019/eccc/En4-350-2-2019-eng.pdf.

54 S. Langlois-Bertrand, K. Vaillancourt., L. Beaumier, M. Pied., O. Bahn, and N. Mousseau, *Canadian Energy Outlook 2021 – Horizon 2060* (Montreal: Institut de l'énergie Trottier and e3c Hub, 2021).

55 Macdonald, Monstadt, and Kern, *Allocating Canadian Greenhouse Gas.*

56 Macdonald, Monstadt, and Kern, *Allocating Canadian Greenhouse Gas*, 1354, fig. 1.

57 Douglas Macdonald, *Carbon Province, Hydro Province: The Challenge of Canadian Energy and Climate Federalism* (Toronto: University of Toronto Press, 2020), 9.

58 Macdonald, *Carbon Province*, 270.

59 Macdonald, *Carbon Province*, 267.

6 Deference: Where Are the Guardians?

1 Stephen Breyer, *Breaking the Vicious Circle: Toward Effective Risk Regulation* (Cambridge, MA: Harvard University Press, 1993).

2 Adrian Vermeule, "The Administrative State: Law, Democracy, and Knowledge," in *The Oxford Handbook of the US Constitution*, ed. Mark Tushnet, Mark A. Graber, and Sanford Levinson (Oxford: Oxford University Press, 2015), 260, 277.

3 Amartya Sen, *Development as Freedom* (New York: Anchor Books, 1999).

4 Vermeule, "Administrative State," 271–2; Jacob E. Gersen and Matthew C. Stephenson, "Over-Accountability," *Journal of Legal Analysis* 6, no. 2 (2014): 187.

5 House of Commons, *The National Research Universal Reactor Shutdown and the Future of Medical Isotope Production and Research in Canada: Report of the Standing Committee on Natural Resources*, November 2010, 9–10, https://www.ourcommons.ca/Content/Committee/403/RNNR/Reports/RP4500827/rnnrrp02/rnnrrp02-e.pdf; Kelly Grant, "Doctors Fear Shortage as Chalk River Reactor Halts Production of Isotope," *Globe and Mail*, 28 October 2016, https://www.theglobeandmail.com/news/national/medical-isotope-shortage-feared-as-chalk-river-reactor-closes/article32582481/.

6 Andrew Green, "Can There Be Too Much Context in Administrative Law?: Setting the Standard of Review in Canadian Administrative Law," *UBC Law Review* 47, no. 2 (2014): 446.

7 *Canada (Minister of Citizenship and Immigration) v Vavilov*, 2019 SCC 65 at para 10 [*Vavilov*] (Can.).

8 *Vavilov* at para 15.

9 There is a sizeable literature on what judges take into account in making decisions, such as the extent to which they decide in accordance with their own political preferences and when they attend to public opinion on an issue. See, for example, Ben Alarie and Andrew Green, *Commitment and Cooperation on High Courts: A Cross-Country Examination of Institutional Constraints on Judges* (Oxford: Oxford University Press, 2017).

10 *Tsleil-Waututh Nation v Canada (Attorney General)*, 2018 FCA 153 at paras 206–27 [*Tsleil-Waututh*] (Can.).

11 For a good discussion of the scope of the Aboriginal and Indigenous law, see Janna Promislow and Naiomi Mettalic, "Realizing Aboriginal Administrative Law," in *Administrative Law in Context*, 3rd ed., ed. Colleen M. Flood and Lorne Sossin (Toronto: Emond, 2018), 87–137.

12 *Clyde River (Hamlet) v Petroleum Geo-Services Inc.*, 2017 SCC 40 at paras 19–24; *Tsleil-Waututh* at para 599.

13 *Gitxaala Nation v Canada*, 2016 FCA 187 at para 182.

14 *United Nations Declaration on the Rights of Indigenous Peoples Act*, S.C. 2021, c. 14.

15 Andrew Green, "On Thin Ice: Meeting Canada's Paris Climate Commitments," *Journal of Environmental Law and Practice* 32, no. 1 (2018): 122.

16 Vermeule, "Administrative State," 260.

17 This idea of disappointment leading to shifts comes from Albert O. Hirschman, *Shifting Involvements: Private Interest and Public Action* (Princeton: Princeton University Press, 2002).

18 Jason MacLean, "Regulatory Capture and the Role of Academics in Public Policymaking: Lessons from Canada's Environmental Regulatory Review Process," *UBC Law Review* 52, no. 2 (2019): 479–552; Jason Maclean, "Striking at the Root Problem of Canadian Environmental Law: Identifying and Escaping Regulatory Capture," *Journal of Environmental Law and Practice* 29 (2016): 111–28; Brendan Haley, "From Staples Trap to Carbon Trap: Canada's Peculiar Form of Carbon Lock-In," *Studies in Political Economy* 88, no. 1 (2011): 97; Brendan Haley, "The Staple Theory and the Carbon Trap," in *The Staple Theory @ 50: Reflections on the Lasting Significance of Mel Watkins' "A Staple Theory of Economic Growth,"* ed. Jim

Stanford (Ottawa: Canadian Centre for Policy Alternatives, 2020), 77,
https://www.policyalternatives.ca/sites/default/files/uploads
/publications/National%20Office/2020/04/Staple%20Theory%20at%2050
%20-%202020%20version.pdf.

19 MacLean, "Regulatory Capture," 502. He goes further in some articles,
noting that "systemic corruption blocks principled reforms and fuels
unprincipled reforms in Canadian environmental law – it is at the root of
every identifiable systemic weakness infecting Canadian environmental
law today, both federally and provincially." MacLean, "Root Problem," 113.

20 David R. Boyd, *Unnatural Law: Rethinking Canadian Environmental Law and
Policy* (Vancouver: UBC Press, 2003).

21 Mark Winfield, *A New Era of Environmental Governance in Canada: Better
Decisions Regarding Infrastructure and Resource Development Projects*, Green
Prosperity Papers (Toronto: Metcalf Foundation, 2016), 7–8, https://
metcalffoundation.com/site/uploads/2016/05/Metcalf_Green
-Prosperity-Papers_Era-of-Governance_final_web.pdf.

22 Jacob Poushter and Christine Huang, *Climate Change Still Seen as the Top
Global Threat, But Cyberattacks a Rising Concern* (Washington, DC: Pew
Research Center, 2019), 3, 11, https://www.pewresearch.org/global
/wp-content/uploads/sites/2/2019/02/Pew-Research-Center_Global
-Threats-2018-Report_2019-02-10.pdf.

23 A 2020 Ipsos poll found "that Canadians may not be as environmentally
conscious" compared to the global average. For example, "two in three
(64%) think in the long term that climate change is as serious of an issue
as COVID-19; this compares to a global average of 71%." Also, "six in ten
(61%) Canadians think that in the economic recovery from COVID-19, it's
important that government actions prioritize climate change." Jennifer
McLeod Macey, "Two-Thirds of Canadians Think, Long Term, Climate
Change Is as Serious of a Problem as Coronavirus," IPSOS, 22 April
2020, https://www.ipsos.com/en-ca/news-and-polls/Two-Thirds-Of
-Canadians-Think--In-The-Long-Term-Climate-Change-Is-As-Serious-Of
-A-Problem-As-Coronavirus.

24 Jason MacLean, "Manufacturing Consent to Climate Inaction: A Case
Study of The Globe and Mail's Pipeline Coverage," *Dalhousie Law Journal*
42, no. 2 (2019): 288.

25 Mark Jaccard, *The Citizen's Guide to Climate Success: Overcoming Myths
That Hinder Progress* (Cambridge: Cambridge University Press, 2020),

49–50; Geoffrey Supran and Naomi Oreskes, "Assessing ExxonMobil's Climate Change Communications (1977–2014)," *Environmental Research Letters* 12, no. 8 (2017); Emily Holden, "How the Oil Industry Has Spent Billions to Control the Climate Change Conversation," *Guardian*, 8 January 2020, https://www.theguardian.com/business/2020 /jan/08/oil-companies-climate-crisis-pr-spending; Amy Westervelt, "How the Fossil Fuel Industry Got the Media to Think Climate Change Was Debatable," *Washington Post*, 10 January 2019, https://www .washingtonpost.com/outlook/2019/01/10/how-fossil-fuel-industry -got-media-think-climate-change-was-debatable/.

26 Jaccard, *Citizen's Guide*.

27 *National Energy Board Act*, R.S.C., 1985, c. N-7, s. 3(1) (Can.) (as repealed by 2019, c. 28, s. 44) provided: "There is hereby established a Board, to be called the National Energy Board, consisting of not more than nine members to be appointed by the Governor in Council" [Cabinet]. These members are appointed for seven years but can be removed by Cabinet ("on address of the Senate and the House of Commons" [ibid., s. 3(2)]).

28 Winfield, *New Era*, 8.

29 Winfield, *New Era*, 8. See also Jaccard, *Citizen's Guide*, chap. 5.

30 ECO, *Good Choices, Bad Choices: Environmental Rights and Environmental Protection in Ontario* (Toronto: Environmental Commissioner of Ontario, 2017), 100–11, 121–41, https://www.auditor.on.ca/en/content /reporttopics/envreports/env17/Good-Choices-Bad-Choices.pdf; "Defending the Rights of Chemical Valley Residents: Charter Challenge," Ecojustice, accessed 28 August 2020, https://ecojustice.ca/case /defending-the-rights-of-chemical-valley-residents-charter-challenge/; Lauren Wortsman, "'Greening' the Charter: Section 7 and the Right to a Healthy Environment," *Dalhousie Journal of Legal Studies* 28 (2019): 245–89.

31 *La Rose et al v. Canada (AG)*, 2020 FC 1008.

32 *Mathur v. Ontario*, 2020 ONSC 6918.

33 Naiomi Metallic, "Deference and Legal Frameworks Not Designed By, For or With Us," *Administrative Law Matters* (blog), 27 February 2018, https://www.administrativelawmatters.com/blog/2018/02/27 /deference-and-legal-frameworks-not-designed-by-for-or-with-us -naiomi-metallic/; Promislow and Mettalic, "Realizing Aboriginal Administrative Law," 116–23; Andrew Green, "Judicial Influence on the Duty to Consult and Accommodate" (forthcoming) *Osgoode Hall Law*

Journal 56, no. 3 (2020): 529–63, https://digitalcommons.osgoode.yorku
.ca/ohlj/vol56/iss3/.

34 For a discussion of the incentives of the court's view of the duty on the
amount of litigation, see Green, "Judicial Influence."

35 Jocelyn Stacey, "The Environmental Emergency and the Legality of
Discretion in Environmental Law," *Osgoode Hall Law Journal* 52, no. 3
(2015): 989.

36 This view, which Stacey calls the "environmental law reform" view, is
held by a wide variety of scholars writing on Canadian environmental.
Stacey, "Environmental Emergency," 997–9. See also Boyd, *Unnatural
Law*; Bruce Pardy, "The Unbearable Licence of Being the Executive:
A Response to Stacey's Permanent Environmental Emergency," *Osgoode
Hall Law Journal* 52, no. 3 (2015): 1045; Bruce Pardy, "In Search of the
Holy Grail of Environmental Law: A Rule to Solve the Problem," *McGill
International Journal of Sustainable Development Law* 1 (2005): 29–58;
Jason MacLean and Chris Tollefson, "Climate-Proofing Judicial Review
after Paris: Judicial Competence, Capacity, and Courage," *Journal of
Environmental Law and Practice* 31, no. 3 (2018): 245–70; Andrew Green,
"Delegation and Consultation: How the Administrative State Functions
and the Importance of Rules," in *Administrative Law in Context*, 3rd ed.,
ed. Colleen M. Flood and Lorne Sossin (Toronto: Emond, 2018), 307–40.

37 Stacey, "Environmental Emergency," 1012–13; Green, "Delegation and
Consultation," 333–9; MacLean and Tollefson, "Climate-Proofing," 249, 252–3.

38 *Vavilov* at para 12.

39 *Vavilov* at para 199 (per Abella and Karakatsanis, JJ [concurring]).

40 *Vavilov* at para 239 (per Abella and Karakatsanis JJ [concurring]).

41 *Vavilov* at para 241.

42 On the general incentives from the choice of standard of review, see
Green, "Too Much Context," 456–9.

43 MacLean, "Root Problem," 113, 125–8.

7 Focusing on People

1 Recognizing the Duty of the Federal Government to Create a Green New
Deal, H.R. Res. 109, 116th Cong. (2019); see also Recognizing the Duty of
the Federal Government to Create a Green New Deal, S. Res. 59,
116th Cong. (2019).

2 "The Leap Manifesto: A Call for a Canada Based on Caring for the Earth and One Another," Leap, accessed 23 June 2020, https://leapmanifesto .org/en/the-leap-manifesto/#manifesto-content.

3 Bruce Yandle, "Bootleggers and Baptists in Retrospect," *Regulation* 22, no. 3 (1999): 5–7, ProQuest.

4 For a discussion of the differences in the ozone and climate contexts, see Cass Sunstein, "Montreal versus Kyoto: A Tale of Two Protocols," *Harvard Environmental Law Rev*iew 31 (2007): 1.

5 David A. Green, W. Craig Riddell, and France St-Hilaire, "Income Inequality in Canada: Driving Forces, Outcomes and Policy," in *The Art of the State*, vol. 5: *Income Inequality: The Canadian Story*, ed. David A. Green, W. Craig Riddell, and France St-Hilaire, (Montreal: Institute for Research on Public Policy, 2016), 1–73, https://irpp.org/wp-content /uploads/2017/02/aots5-intro.pdf.

6 The Gini after-tax income continued to be in the 0.31 range between 2014 and 2018. "Table 11–10–0134–01: Gini Coefficients of Adjusted Market, Total and After-Tax Income," Statistics Canada, last modified 6 August 2020, https://doi.org/10.25318/1110013401-eng.

7 Green, Riddell, and St-Hilaire, "Income Inequality."

8 "Median After-Tax Income, Canada and Provinces, 2014–2018," Statistics Canada, last modified 24 February 2020, https://www150.statcan.gc.ca /n1/daily-quotidien/200224/t003a-eng.htm.

9 Green, Riddell, and St-Hilaire, "Income Inequality," 9.

10 "Labour Force Survey, May 2020," Statistics Canada, 5 June 2020, https://www150.statcan.gc.ca/n1/daily-quotidien/200605/dq200605a -eng.htm?HPA=1&indid=3587-2&indgeo=0; "Gross Domestic Product, Income and Expenditure, First Quarter 2020," Statistics Canada, 29 May 2020, https://www150.statcan.gc.ca/n1/daily-quotidien/200529 /dq200529a-eng.htm?HPA=1&indid=3278-1&indgeo=0.

11 Derek Messacar, René Morissette, and Zechuan Deng, *Inequality in the Feasibility of Working from Home During and After COVID-19*, (Ottawa: Statistics Canada, 2020), https://www150.statcan.gc.ca/n1/pub/45-28 -0001/2020001/article/00029-eng.pdf; Canadian Women's Foundation et al., *Resetting Normal: Women, Decent Work and Canada's Fractured Care Economy* (Toronto: Canadian Women's Foundation, 2020), https://www .policyalternatives.ca/sites/default/files/uploads/publications /National%20Office/2020/07/ResettingNormal-Women-Decent-Work

-and-Care.pdf; Feng Hou, Kristyn Frank, and Christoph Schimmele, *Economic Impact of COVID-19 among Visible Minority Groups*, StatCan COVID-19: Data to Insights for a Better Canada (Ottawa: Statistics Canada, 2020), https://www150.statcan.gc.ca/n1/en/pub/45-28 -0001/2020001/article/00042-eng.pdf?st=E0VsY8_B.

12 Dale Beugin et al., *Provincial Carbon Pricing and Household Fairness* (Montreal: Canada's Ecofiscal Commission, 2016), 5, http://ecofiscal .ca/wp-content/uploads/2016/04/Ecofiscal-Commission-Provincial -Carbon-Pricing-Household-Fairness-Report-April-2016.pdf; Parliamentary Budgetary Officer, *Fiscal and Distributional Analysis of the Federal Carbon Pricing System* (Ottawa: PBO, 2019), 9, https://www.pbo -dpb.gc.ca/web/default/files/Documents/Reports/2019/Federal %20Carbon/Federal_carbon_pricing_EN.pdf.

13 Beugin et al., *Household Fairness*, 5 (noting that some firms may not pass some or even all of the carbon costs onto consumers, depending on the market they face). See also Alexander R. Barron, Marc A.C. Hafstead, and Adele C. Morris, "Policy Insights from Comparing Carbon Pricing Modeling Scenarios," Climate and Energy Economics Discussion Paper, Environmental Science and Policy: Faculty Publications, Smith College, Northampton, MA, 2019, 6, 8–9, https://scholarworks.smith.edu /env_facpubs/6.

14 Justin Caron et al., "Distributional Implications of a National CO2 Tax in the US across Income Classes and Regions: A Multi-model Overview," *Climate Change Economics* 9, no. 1 (2018): 22, 25, https://doi.org/10.1142 /S2010007818400043 (arguing the carbon intensity of industry can be reflected in variations in regional cost incidence of carbon pricing).

15 Canada's Ecofiscal Commission, *Choose Wisely: Options and Trade-Offs in Recycling Carbon Pricing Revenues* (Montreal: Canada's Ecofiscal Commission, 2016), 8, http://ecofiscal.ca/wp-content/uploads /2016/04/Ecofiscal-Commission-Choose-Wisely-Carbon-Pricing- Revenue-Recycling-Report-April-2016.pdf; Parliamentary Budgetary Officer, *Fiscal and Distributional Analysis*, 9. See, more generally, Barron, Hafstead, and Morris, "Policy Insights," 9, 11 (examining effects of carbon pricing across eleven CGE models and finding geographic variation in costs depends on the carbon-intensity of the regional economy).

16 Parliamentary Budgetary Officer, *Fiscal and Distributional Analysis*, 9–10.

17 Parliamentary Budget Officer, *Reviewing the Fiscal and Distributional Analysis of the Federal Carbon Pricing System* (Ottawa: PBO, 2020), 8, https://www.pbo-dpb.gc.ca/web/default/files/Documents/Reports/RP-1920-024-S/RP-1920-024-S_en.pdf.
18 Marisa Beck et al., "Carbon Tax and Revenue Recycling: Impacts on Households in British Columbia," *Resource and Energy Economics* 41 (2015): 60, http://dx.doi.org/10.1016/j.reseneeco.2015.04.005 (using a CGE model to estimate the effects of the BC carbon tax on households).
19 Beugin et al., *Household Fairness*, 4.
20 Sheila Block, "Canada's Population Is Changing but Income Inequality Remains a Problem," *Behind the Numbers* (blog), Canadian Centre for Policy Alternatives, 27 October 2017, http://behindthenumbers.ca/2017/10/27/population-changing-income-inequality-remains/ (looking at the 2016 Census numbers). See also "Household Income in Canada: Key Results from the 2016 Census," Statistics Canada, 13 September 2017, https://www150.statcan.gc.ca/n1/daily-quotidien/170913/dq170913a-eng.htm; OECD, *Policies for Stronger and More Inclusive Growth in Canada*, Better Policies Series (Paris: OECD, 2017), 3–6, 12, https://doi.org/10.1787/9789264277946-en.
21 Green, Riddell, and St-Hilaire, "Income Inequality," 19–20.
22 See also Sheila Block, Grace-Edward Galabuzi, and Ricardo Tranjan, *Canada's Colour Coded Income Inequality* (Ottawa: Canadian Centre for Policy Alternatives, 2019), 12–3, https://www.policyalternatives.ca/sites/default/files/uploads/publications/National%20Office/2019/12/Canada%27s%20Colour%20Coded%20Income%20Inequality.pdf (finding in 2015 the average employment income for men [non-racialized] was 56,920; whereas the average employment income for men [all racialized groups] was 44,423).
23 Natural Resources Canada, *Energy Fact Book: 2019–2020* (Ottawa: Natural Resources Canada, 2020), 13, https://www.nrcan.gc.ca/sites/www.nrcan.gc.ca/files/energy/energy-factbook_EN-feb14-2020.pdf (the sector includes oil and gas extraction, exploration and crude oil pipelines).
24 Canadian Association of Petroleum Producers, "Canadian Economic Contribution," accessed 7 August 2020, https://www.capp.ca/economy/canadian-economic-contribution/.
25 Task Force on Just Transition for Canadian Coal Power Workers and Communities, *A Just and Fair Transition for Canadian Coal Power Workers*

and Communities (Gatineau: ECCC, 2019), vii, http://publications.gc.ca /collections/collection_2019/eccc/En4-361-2019-eng.pdf.

26 "CAPP Issues Statement on Historic Low Oil Prices," Canadian Association of Petroleum Producers, 20 April 2020, https://www.capp .ca/news-releases/capp-issues-statement-on-historic-low-oil-prices/; "Canadian Energy Companies Cenovus and Husky Swing to Huge Losses Amid Oil Price Collapse," CBC News, 29 April 2020, https:// www.cbc.ca/news/business/oil-patch-earnings-wednesday-1.5548838.

27 Tony Seskus, "Number of Canadians Employed by Oil and Gas Sector Falls by 14,000, Data Shows," CBC News, 19 June 2020, https://www .cbc.ca/news/business/canadian-oil-and-gas-jobs-1.5619621.

28 David A. Green et al., "Economy Wide Spillovers from Booms: Long-Distance Commuting and the Spread of Wage Effects," *Journal of Labor Economics* 37, no. S2 (2019): S676–S680, https://doi.org/10.1086/703362.

29 Beugin et al., *Household Fairness*

30 Akio Yamazaki, "Jobs and Climate Policy: Evidence from British Columbia's Revenue-Neutral Carbon Tax," *Journal of Environmental Economics and Management* 83 (May 2017): 198, https://doi.org/10.1016 /j.jeem.2017.03.003.

31 Chi Man Yip, "On the Labor Market Consequences of Environmental Taxes," *Journal of Environmental Economics and Management* 89 (May 2018): 137, https://doi.org/10.1016/j.jeem.2018.03.004.

32 This discussion is based in part on Andrew Green, "On Thin Ice: Meeting Canada's Paris Climate Commitments," *Journal of Environmental Law and Practice* 32, no. 1 (2018): 104, ProQuest.

33 John Rawls, *A Theory of Justice* (London: Harvard University Press, 2009 [1971]), 3–53.

34 On the connection between Rawls and welfare economics, see David A. Green, "What Is a Minimum Wage For? Empirical Results and Theories of Justice," *Canadian Public Policy/Analyse de Politiques* 40, no. 4 (2014): 293–314, www.jstor.org/stable/24365133.

35 Amartya Sen, *Development as Freedom* (New York: Anchor Books, 2000), 18.

36 Sen, *Development as Freedom*, 295.

37 Sandeep Pai, Kathryn Harrison, and Hisham Zerriffi, "A Systemic Review of the Key Elements of a Just Transition for Fossil Fuel Workers," Clean Economy Working Paper Series WP 20–04, Smart Prosperity Institute, University of Ottawa, April 2020, https://institute

.smartprosperity.ca/sites/default/files/transitionforfossilfuelworkers.pdf
(summarizing the literature on what makes a just transition).

38 David J. Doorey, "Just Transitions Law: Putting Labour Law to Work
on Climate Change," *Journal of Environmental Law and Practice* 30, no. 2
(2017): 223, ProQuest.

39 Amartya Sen, *The Idea of Justice* (Cambridge, MA: Belknap, 2009), 26.

40 Sen notes, "Freedoms are not only the primary ends of development,
they are also among its principal means." Sen, *Development as Freedom*,
10. See also Pai, Harrison, and Zerriffi, "Systemic Review," 4 (noting that
some base their vision of a just transition on the pragmatic notion that it
is the only way to bring about change).

41 Parliamentary Budget Officer, *Reviewing the Fiscal and Distributional
Analysis*, 3.

42 Beck et al., "Impacts on Households," 43, 57–8, 60; Mark Jaccard, Mikela
Hein and Tiffany Vass, *Is Win-Win Possible? Can Canada's Government
Achieve Its Paris Commitment … and Get Re-elected?* (Vancouver: School
of Resource and Environmental Management, Simon Fraser University,
2016), 6, http://www.sfu.ca/content/dam/sfu/emrg/Publications
/Research_Publications/2016%20Jaccard,%20Hein%20&%20Vass%20
-%20Is%20Win-Win%20Possible.pdf.

43 Mark Jaccard, *The Citizen's Guide to Climate Success: Overcoming Myths
That Hinder Progress* (Cambridge: Cambridge University Press, 2020),
109–27, doi.org/10.1017/9781108783453; Jaccard, Hein, and Vass, *Is Win-
Win Possible?*, 4.

44 See Mark Winfield and Abdeali Saherwala, "Phasing-Out Coal-Fired
Electricity in Ontario," January 2021, https://sei.info.yorku.ca/files
/2021/01/Coal-Phase-Out-January-2021.pdf?x60126.

45 ECCC, *A Healthy Environment and a Healthy Economy* (Gatineau: ECCC,
2020), https://www.canada.ca/content/dam/eccc/documents/pdf
/climate-change/climate-plan/healthy_environment_healthy_economy
_plan.pdf.

46 Canadian Institute for Climate Choices, *Sink or Swim: Transforming
Canada's Economy for a Global Low-Carbon Future* (Ottawa: CICC, 2021),
https://climatechoices.ca/wp-content/uploads/2021/10/CICC-Sink-or
-Swim-English-Final-High-Res.pdf.

47 Green, Riddell and St-Hilaire, "Income Inequality."

48 Western Economic Diversification Canada, "Government of Canada Supports a Just and Fair Coal Energy Transition for Alberta," *Cision*, 3 September 2019, https://www.newswire.ca/news-releases /government-of-canada-supports-a-just-and-fair-coal-energy-transition -for-alberta-808745697.html; Task Force on Just Transition for Canadian Coal Power Workers and Communities, *Just and Fair Transition*, vii–ix.

49 Charlie Pinkerton, "First Environmental Action for Liberals Could Include 'Just Transition Act,' Strengthening CEPA," *iPolitics*, 21 January 2020, https://ipolitics.ca/2020/01/21/first-environmental-action-for -liberals-could-include-just-transition-act-strengthening-cepa/.

50 ECCC, *Healthy Environment, Healthy Economy*, 49–50.

51 For example, see Brendan Haley, "Economy and Climate Need More than Stimulus after COVID-19," *Policy Options*, 27 April 2020, https:// policyoptions.irpp.org/magazines/april-2020/economy-and-climate -need-more-than-stimulus-after-covid-19/; Stephen Cornish and Karel Mayrand, "Canada Needs a 'Green Recovery' to Confront COVID-19 and Other Crises," *Policy Options*, 14 May 2020, https://policyoptions.irpp .org/magazines/may-2020/canada-needs-a-green-recovery-to-confront -covid-19-and-other-crises/; John McNally, "Green Stimulus Offers Canada a Way Forward for Escaping the Next Recession," Smart Prosperity Institute, 26 March 2020, https://institute.smartprosperity.ca /content/green-stimulus-offers-canada-way-forward-escaping-next -recession.

52 Stewart Elgie and John McNally, "3 Ingredients for Smart Stimulus," Smart Prosperity Institute, 30 April 2020, https://institute.smart prosperity.ca/3ingredients. See also Canadian Institute for Climate Choices, *Sink or Swim*; Pembina Institute, *Green Stimulus: Pembina Institute Principles and Recommendations for a 2020 Economic Stimulus Package*, March 2020, 1–2, https://www.pembina.org/reports/green-stimulus.pdf.

53 Scott Barrett, *Why Cooperate? The Incentive to Supply Global Public Goods* (New York: Oxford University Press, 2007), chap. 3, https://doi.org /10.1093/acprof:oso/9780199211890.001.0001.

54 US, HR Res 109, *Recognizing the duty of the Federal Government to create a Green New Deal*, 116th Congress, 2019, https://www.congress.gov/bill /116th-congress/house-resolution/109/text [US, HR Res 109]; see also US, S Res 59, *A resolution recognizing the duty of the Federal Government to*

create a Green New Deal, 116ᵗʰ Congress, 2019; Leap, "The Leap Manifesto: A Call for a Canada Based on Caring for the Earth and One Another," accessed 26 August 2019, https://leapmanifesto.org/en/the-leap -manifesto/; "The Pact for a Green New Deal," accessed 26 August 2019, https://greennewdealcanada.ca; "A Green New Deal for Canada: What It Means," CBC News, 10 May 2019, https://www.cbc.ca/news /technology/what-on-earth-newsletter-green-new-deal-canada-1.5129864.

55 Thomas Piketty and Antoine Vauchez, "Manifesto for the Democratization of Europe," *Social Europe* (newsletter), 11 December 2018, https://www.socialeurope.eu/manifesto-for-the-democratization -of-europe.

56 European Commission, "Financing the Green Transition: The European Green Deal Investment Plan and Just Transition Mechanism," press release no. IP/20/17, 14 January 2020, https://ec.europa.eu/commission /presscorner/detail/en/ip_20_17.

57 International Labour Organization, "Just Transition Commitments Made by Many Countries at UN Climate Action Summit," press release, 23 September 2019, https://www.ilo.org/global/about-the-ilo /newsroom/news/WCMS_721144/lang-en/index.htm; UNFCCC Secretariat, *Just Transition of the Workforce, and the Creation of Decent Work and Quality Jobs* (Geneva: United Nations Framework Convention on Climate Change, 2020), https://unfccc.int/sites/default/files/resource /Just%20transition_for%20posting.pdf; Just Transition Research Collaborative, *Mapping Just Transition(s) to a Low-Carbon World* (Geneva: United Nations Research Institute for Social Development, 2018), http:// www.unrisd.org/unrisd/website/document.nsf/(httpPublications) /9B3F4F10301092C7C12583530035C2A5?OpenDocument.

58 Jaccard, *Citizen's Guide*, 238.

59 This section is based on Andrew Green, "The Evolution of Government as Risk Manager in Canada," 27 April 2004, Microsoft Word document.

60 Pai, Harrison, and Zerriffi, "Systemic Review," 27–8.

61 Hadrian Mertins-Kirkwood and Zaee Deshpande, *Who Is Included in a Just Transition? Considering Social Equity in Canada's Shift to a Zero-Carbon Economy* (Canadian Centre for Policy Alternatives/ACW, 2019), 6.

62 Pai, Harrison, and Zerriffi, "Systemic Review," 23–4 (noting identity as theme in the literature on just transitions).

63 Green, Riddell, and St-Hilaire, "Income Inequality," 33–5.

64 Jaccard call this a "Faustian dilemma." Jaccard, *Citizen's Guide*, 234.
65 Pai, Harrison, and Zerriffi, "Systemic Review," 4–5.

8 Strengthening the National Community

1 Mark Jaccard, *The Citizen's Guide to Climate Success: Overcoming Myths That Hinder Progress* (Cambridge: Cambridge University Press, 2020), 235.
2 Robin Boadway and Benjamin Dachis, *Drilling Down on Royalties: How Canadian Provinces Can Improve Non-Renewable Resource Taxes*, Commentary No. 435 (Toronto: C.D. Howe Institute, 2015), 8, https://www.cdhowe.org/sites/default/files/attachments/research_papers/mixed/Commentary_435_0.pdf.
3 Natural Resources Canada, *Energy Fact Book: 2019–2020* (Natural Resources Canada, July 2019), 10.
4 Trevor Tombe, "Alberta's Long-Term Fiscal Future," *School of Public Policy Publications* 11, no.31 (2018): 7, https://doi.org/10.11575/sppp.v11i0.52965.
5 Blue Ribbon Panel on Alberta's Finances, *Report and Recommendations*, August 2019, 13, https://open.alberta.ca/dataset/081ba74d-95c8-43ab-9097-cef17a9fb59c/resource/257f040a-2645-49e7-b40b-462e4b5c059c/download/blue-ribbon-panel-report.pdf; Konrad Yakabuski, "Under Kenney's Plan, Quebec's Equalization Take Would Grow," *Globe and Mail*, 26 November 2019, https://www.theglobeandmail.com/business/commentary/article-under-kenneys-plan-quebecs-equalization-take-would-grow/; "If Alberta Taxed Like Other Provinces, It Would Have a Huge Budget Surplus," editorial, *Globe and Mail*, 4 March 2020, https://www.theglobeandmail.com/opinion/editorials/article-if-alberta-taxed-like-other-provinces-it-would-have-a-huge-budget/.
6 RBC Economics, *Alberta Announces "Sweeping Actions": $10 B Fiscal Stimulus*, June 2020, https://royal-bank-of-canada-2124.docs.contently.com/v/alberta-announces-sweeping-actions-10-b-fiscal-stimulus?utm_medium=internal&utm_source=website&utm_campaign=fiscal.
7 Sarah Offin, "Alberta Budget Benefits from Oil Prices No One Had Bargained For," 24 June 2021), Global News, https://globalnews.ca/news/7976953/alberta-budget-benefits-from-oil-prices-no-one-had-bargained-for/.
8 Trevor Tombe, "'Final and Unalterable' – But Up for Negotiation: Federal-Provincial Transfers in Canada," *Canadian Tax Journal* 66,

no. 4 (2018): 901, https://www.trevortombe.com/files/tombe_final
_unalterable_2019.pdf.

9 Bruce Campbell, *The Petro-Path Not Taken: Comparing Norway with Canada
   and Alberta's Management of Petroleum Wealth* (Ottawa: Canadian Centre
   for Policy Alternatives, 2013), 44.

10 British Columbia Ministry of Finance, *A Stronger BC, for Everyone – Budget
   2020: A Balanced Plan to Keep BC Moving Forward: Budget and Fiscal Plan
   2020/21–2022/23*, 18 February 2020, 28, table 1.3, https://www.bcbudget
   .gov.bc.ca/2020/pdf/2020_budget_and_fiscal_plan.pdf.

11 Parliamentary Budget Officer, *Canada's Greenhouse Gas Emissions;
   Developments, Prospects and Reductions* (Ottawa: PBO, 2016), 26, https://
   www.pbo-dpb.gc.ca/web/default/files/Documents/Reports/2016
   /ClimateChange/PBO_Climate_Change_EN.pdf.

12 Financial Accountability Office of Ontario, *Cap and Trade: A Financial
   Review of the Decision to Cancel the Cap and Trade Program* (Toronto: FAO,
   2018), 9–10, https://www.fao-on.org/web/default/files/publications
   /ending%20cap%20and%20trade%20oct%202018/Cap%20and%20Trade
   .pdf (although some of the loss of revenue is offset by cancelling some
   related spending programs).

13 Tracy Snoddon and Trevor Tombe, "Analysis of Carbon Tax Treatment in
   Canada's Equalization Program," *Canadian Public Policy* 45, no. 3 (2019):
   379, https://doi.org/10.3138/cpp.2019-036.

14 See, for example, Stephen Ornes et al., "Core Concept: How Does Climate
   Change Influence Extreme Weather? Impact Attribution Research Seeks
   Answers," *Proceedings of the National Academy of Sciences*, 14 August 2018,
   https://www.pnas.org/content/115/33/8232.

15 Canadian Institute for Climate Choices, *The Tip of the Iceberg: Navigating
   the Known and Unknown Costs of Climate Change for Canada* (Ottawa: CICC,
   2020), https://climatechoices.ca/wp-content/uploads/2020/12/Tip-of
   -the-Iceberg-_-CoCC_-Institute_-Full.pdf.

16 Howard Frumkin et al., "Climate Change: The Public Health Response,"
   American Journal of Public Health 98, no. 3 (2008): 435–45, https://doi.org
   /10.2105/ajph.2007.119362; World Health Organization, *COP24 Special
   Report: Health & Climate Change* (Geneva: WHO, 2018), 52, https://www
   .who.int/publications/i/item/cop24-special-report-health-climate
   -change; Jessica Boyle, Maxine Cunningham, and Julie Dekens, *Climate
   Change Adaptation and Canadian Infrastructure: A Review of the Literature*

(Winnipeg: IISD, 2013), 1, https://www.iisd.org/sites/default/files
/publications/adaptation_can_infrastructure.pdf.

17 Auditor General of Canada, *Perspectives on Climate Change Action in
Canada: A Collaborative Report from Auditors General* (Ottawa: Auditor
General of Canada, 2018), 5, http://publications.gc.ca/collections
/collection_2018/bvg-oag/FA3-137-2018-eng.pdf. See also F. Warren
and N. Lulham, eds., *Canada in a Changing Climate: National Issues Report*
(Ottawa: Government of Canada, 2021), https://changingclimate.ca
/national-issues/.

18 Canadian Institute for Climate Choices, *Tip of the Iceberg*, iii.

19 "Disaster Financial Assistance Arrangements (DFAA)," Public Safety
Canada, last modified 24 March 2020, https://www.publicsafety.gc.ca
/cnt/mrgnc-mngmnt/rcvr-dsstrs/dsstr-fnncl-ssstnc-rrngmnts/index
-en.aspx.

20 Parliamentary Budget Officer, *Estimate of the Average Annual Cost for
Disaster Financial Assistance Arrangements Due to Weather Events* (Ottawa:
PBO, 2016), 2, http://www.pbo-dpb.gc.ca/web/default/files/Documents
/Reports/2016/DFAA/DFAA_EN.pdf.

21 Insurance Bureau of Canada and Federation of Canadian Municipalities,
Investing in Canada's Future: The Cost of Climate Adaptation, September
2019, 4, http://assets.ibc.ca/Documents/Disaster/The-Cost-of-Climate
-Adaptation-Summary-EN.pdf.

22 Clean Energy Canada, "Media Brief: The Link between Climate
Change and the Health of Canadians," 20 September 20, 2019, https://
cleanenergycanada.org/the-link-between-climate-change-and-the
-health-of-canadians/.

23 Robin Boadway, "Natural Resource Shocks and the Federal System:
Boon and Curse?" *Fiscal Federalism and the Future of Canada* (conference
proceedings), Queen's Institute of Intergovernmental Relations, 28–29
September 2006, https://www.queensu.ca/iigr/sites/webpublish
.queensu.ca.iigrwww/files/files/WorkingPapers/fiscalImb
/boadway.pdf.

24 Laura Merrill and Franziska Funke, "All Change and No Change: G20
Commitment on Fossil Fuel Subsidy Reform, Ten Years On," *Subsidy
Watch Blog*, IISD, 8 October 2019, https://www.iisd.org/gsi/subsidy
-watch-blog/all-change-and-no-change-g20-commitment-fossil-fuel
-subsidy-reform-ten-years.

25 International Institute for Sustainable Development, "Unpacking Canada's Fossil Fuel Subsidies: Their Size, Impacts and Why They Must Go," accessed 17 August 2020, https://www.iisd.org/faq/unpacking -canadas-fossil-fuel-subsidies/.

26 Shelagh Whitley et al., *G7 Fossil Fuel Subsidy Scorecard: Tracking the Phase-Out of Fiscal Support and Public Finance for Oil, Gas and Coal* (London: ODI, 2018), 16, https://www.odi.org/sites/odi.org.uk/files/resource -documents/12222.pdf.

27 Yanick Touchette and Philip Gass, *Public Cash for Oil and Gas: Mapping Federal Fiscal Support for Fossil Fuels* (Winnipeg: IISD, 2018), iii, https:// www.iisd.org/sites/default/files/publications/public-cash-oil-gas-en.pdf.

28 Liberal Party of Canada, *Real Change: A New Plan for A Strong Middle Class* (Liberal Party of Canada, 2015), 40, https://www.liberal.ca/wp-content /uploads/2015/10/New-plan-for-a-strong-middle-class.pdf.

29 Auditor General of Canada, *Reports of the Commissioner of the Environment and Sustainable Development to the Parliament of Canada: Report 3: Tax Subsidies for Fossil Fuels: Department of Finance Canada: Independent Auditor's Report* (Ottawa: Auditor General of Canada, 2019), 3, http:// publications.gc.ca/collections/collection_2019/bvg-oag/FA1-26-2019-1 -3-eng.pdf.

30 Mitchell Beer, "It's True: The Oil Patch Doesn't Need Handouts," *Policy Options*, 21 January 2019, https://policyoptions.irpp.org/magazines /january-2019/its-true-the-oil-patch-doesnt-need-handouts/.

31 "Backgrounder: Trans Mountain Expansion Project Construction Accelerating," Trans Mountain, 7 February 2020, https://www .transmountain.com/news/2020/backgrounder-trans-mountain -expansion-project-construction-accelerating. See also Department of Finance Canada, "Building the Trans Mountain Expansion Project," last modified 7 February 2020, https://www.canada.ca/en/department -finance/news/2020/02/building-the-trans-mountain-expansion -project.html.

32 Government of Canada, "Canada's COVID-19 Economic Response Plan," last modified 17 August 2020, https://www.canada.ca/en/department -finance/economic-response-plan.html; Department of Finance Canada, "Canada's COVID-19 Economic Response Plan: New Support to Protect Canadian Jobs," last modified 17 April 2020, https://www.canada.ca /en/department-finance/news/2020/04/canadas-covid-19-economic

-response-plan-new-support-to-protect-canadian-jobs.html; Nigel Bankes et al., "Governance and Accountability: Preconditions for Committing Public Funds to Orphan Wells and Facilities and Inactive Wells," *ABlawg: The University of Calgary Faculty of Law Blog*, 24 April 2020, https://ablawg.ca/2020/04/24/governance-and-accountability-preconditions-for-committing-public-funds-to-orphan-wells-and-facilities-and-inactive-wells/; Canadian Association of Petroleum Producers, "Issues Statement"; Canada Development Investment Corporation, "Large Employer Emergency Financing Facility Factsheet," last modified 10 July 2020, https://www.cdev.gc.ca/leeff-factsheet/.

33 International Institute for Sustainable Development, *Pipelines or Progress: Government Support for Oil and Gas Pipelines in Canada* (Winnipeg: IISD, 5 July 2021), https://www.iisd.org/publications/oil-gas-pipelines-green-recovery-canada.

34 Vanessa Corkal, Julia Levin, and Philip Gass, *Canada's Federal Fossil Fuel Subsidies in 2020* (Winnipeg: IISD, 2020), 1, https://www.iisd.org/sites/default/files/publications/canada-fossil-fuel-subsidies-2020-en.pdf.

35 Auditor General of Canada, *Report 3: Tax Subsidies*, 5; Auditor General of Canada, *Reports of the Commissioner of the Environment and Sustainable Development to the Parliament of Canada: Report 4: Non-Tax Subsidies for Fossil Fuels: Environment and Climate Change Canada: Independent Auditor's Report* (Ottawa: Auditor General of Canada, 2019), 4, http://publications.gc.ca/collections/collection_2019/bvg-oag/FA1-26-2019-1-4-eng.pdf. Finance Canada replied that it "has now phased out most tax preferences for fossil fuel production." Auditor General of Canada, *Report 3: Tax Subsidies*, 13. See also Auditor General of Canada, *Report 4: Non-Tax Subsidies*, 14, and Jolson Lim, "Watchdog, Finance Canada Still at Odds Over 'Inefficient' Oil and Gas Subsidies," *iPolitics*, 10 June 2019, https://ipolitics.ca/2019/06/10/watchdog-finance-canada-still-at-odds-over-inefficient-oil-and-gas-subsidies/.

36 International Institute for Sustainable Development, *Federal Fossil Fuel Subsidies in Canada: COVID-19 Edition* (Winnipeg: IISD, 25 February 2021), https://www.iisd.org/publications/fossil-fuel-subsidies-canada-covid-19; Environmental Defence, *Paying Polluters: Federal Financial Support to Oil and Gas in 2020* (Toronto: EDC, April 2021), https://environmentaldefence.ca/wp-content/uploads/2021/04/Federal-FossilFuelSubsidies-April-2021.pdf.

37 Environmental Defence Canada and International Institute for Sustainable Development, *Doubling Down with Taxpayer Dollars: Fossil Fuel Subsidies from the Alberta Government* (Winnipeg/Toronto, EDC/IISD, February 2019), 3, 7, https://d36rd3gki5z3d3.cloudfront.net/wp-content/uploads/2019/02/EDC_IISD_AlbertaFFSReportFINAL.pdf?x99135.

38 Vanessa Corkal and Philip Gass, *The (Public) Cost of Pollution: Ontario's Fossil Fuel Subsidies* (Winnipeg: IISD, 2019), 1, https://www.iisd.org/sites/default/files/publications/public-cost-of-pollution.pdf. See also IISD, "Unpacking."

39 Vanessa Corkal and Philip Gass, *Locked In and Losing Out: British Columbia's Fossil Fuel Subsidies* (Winnipeg: IISD, 2019), iii, https://www.iisd.org/sites/default/files/publications/locked-in-losing-out.pdf.

40 Stockholm Environment Institute et al., *The Production Gap: The Discrepancy between Countries' Planned Fossil Fuel Production and Global Production Levels Consistent with Limiting Warming to 1.5°C or 2°C* (2019), https://wedocs.unep.org/bitstream/handle/20.500.11822/30822/PGR19.pdf?sequence=1&isAllowed=y.

41 Damian Carrington, "'Reckless': G20 States Subsidized Fuels by $3tn since 2015 Says Report," *The Guardian*, 20 July 2021, https://www.theguardian.com/environment/2021/jul/20/g20-states-subsidised-fossil-fuels-2015-coal-oil-gas-cliamte-crisis.

42 Merrill and Funke, "All Change." See also Peter Erickson et al., "Effect of Subsidies to Fossil Fuel Companies on United States Crude Oil Production," *Nature Energy* 2 (2017): 891–8, https://doi.org/10.1038/s41560-017-0009-8.

43 ECCC, *A Healthy Environment and a Healthy Economy* (Gatineau: ECCC, 2020), 37, https://www.canada.ca/content/dam/eccc/documents/pdf/climate-change/climate-plan/healthy_environment_healthy_economy_plan.pdf.

44 Tombe, "Up for Negotiation," 881–4. See also James P. Feehan, *Canada's Equalization Program: Political Debates and Opportunities for Reform*, IRPP Insight no. 30 (Montreal: Institute for Research on Public Policy, 2020), 6–7, https://irpp.org/wp-content/uploads/2020/01/Canada's-Equalization-Program-Political-Debates-and-Opportunities-for-Reform.pdf.

45 As an example, Tombe points out that a province with more high-income earners will be able to raise more money from a 10% tax than a province with fewer. Tombe, "Up for Negotiation," 886.

46 Government of Canada, "Major Federal Transfers," last modified 2 February 2017, https://www.canada.ca/en/department-finance /programs/federal-transfers/major-federal-transfers.html#Quebec, although if we look on a per capita basis, Quebec had the lowest payments; Feehan, *Canada's Equalization Program*, 6.

47 Boadway, "Natural Resource Shocks," 8–11. See also Tombe, "Up for Negotiation," 914.

48 Daniel Béland et al., "The Challenge for Canada's Equalization Program," *Policy Options*, 18 July 2018, https://policyoptions.irpp.org /magazines/july-2018/challenge-canadas-equalization-program/. Tombe argues, however, that "provinces that receive equalization will receive less if they develop their resources, and although only 50 percent of resource revenues are included, the fiscal capacity cap almost entirely eliminates the incentive to develop natural resources in provinces where it binds." Tombe, "Up for Negotiation," 914.

49 Feehan, *Canada's Equalization Program*, 4, 11.

50 Fair Deal Panel, *Report to Government*, May 2020, 7, https://open.alberta .ca/dataset/d8933f27-5f81-4cbb-97c1-f56b45b09a74/resource/d5836820 -d81f-4042-b24e-b04e012f4cde/download/fair-deal-panel-report-to -government-may-2020.pdf.

51 "Albertans Support Bid to Change Equalization, Narrowly Turn Down Year-Round Daylight Time," CBC, 26 October 2021, https://www.cbc .ca/news/canada/edmonton/referendum-alberta-equalization-daylight -time-senate-1.6225309.

52 Bev Dahlby, "Reforming the Federal Fiscal Stabilization Program," *School of Public Policy Publications* 12, no. 18 (2019): 8–9, http://dx.doi.org /10.11575/sppp.v12i0.68076.

53 Trevor Tombe, *An (Overdue) Review of Canada's Fiscal Stabilization Program*, IRPP Insight no. 31 (Montreal: IRPP, 2020), 8, https://irpp.org/wp -content/uploads/2020/02/An-Overdue-Review-of-Canada's-Fiscal -Stabilization-Program.pdf.

54 Dahlby, "Reforming," 3–4.

55 Dahlby, "Reforming," 5–7, 15.

56 Tombe, *An (Overdue) Review,* 4.
57 Tombe, *An (Overdue) Review,* 14–24.
58 Tombe, "Up for Negotiation," 905; Béland et al., "The Challenge."
59 Robert Mansell, Mukesh Khanal, and Trevor Tombe, "The Regional Distribution of Federal Fiscal Balances: Who Pays, Who Gets and Why It Matters," *School of Public Policy Publications* 13, no. 14 (2020): 2, https://doi.org/10.11575/sppp.v13i0.69872.
60 Mansell, Khanal, and Tombe, "Regional Distribution," 1–2.
61 Tombe, "Up for Negotiation," 910.
62 Tombe, "Up for Negotiation," 872.
63 Mansell, Khanal, and Tombe, "Regional Distribution," 28.
64 Boadway, "Natural Resource Shocks," 10.
65 Mansell, Khanal, and Tombe, "Regional Distribution," 28.
66 Conference Board of Canada, "Uneven Provincial Economic Performance in Store for 2020," 26 November 2019, https://www.globenewswire.com/fr/news-release/2019/11/26/1952781/0/en/Uneven-provincial-economic-performance-in-store-for-2020.html.
67 Trevor Tombe, "Opinion: Political Games Won't Solve Alberta's Problems," CBC News, 23 June 2020, https://www.cbc.ca/news/canada/calgary/alberta-fair-deal-panel-trevor-tombe-1.5622060.
68 Snoddon and Tombe, "Carbon Tax Treatment," 379, 388.
69 Tombe, "Alberta's Long-Term Fiscal Future," 27; Tombe, *An (Overdue) Review,* 5, 22.
70 Tombe, *An (Overdue) Review,* 4–5, 7.
71 See Tombe, *An (Overdue) Review,* 23.
72 Seth Klein, *A Good War: Mobilizing Canada for the Climate Emergency* (Vancouver: ECW Press, 2020).
73 Trevor Tombe, "Don't Blame Equalization for Alberta's Fiscal Mess," *Globe and Mail,* 21 October 2017, https://www.theglobeandmail.com/opinion/dont-blame-equalization-for-albertas-fiscal-mess/article36680619/.

9 Cultivating Cooperation

1 Albert O. Hirschman, *Exit, Voice and Loyalty: Responses to Decline in Firms, Organizations and States* (Cambridge, MA: Harvard University Press, 1970), 1–43.

2 *114957 Canada Ltée (Spray Tech, Société d'arrosage) v. Hudson (Town)*, 2001 SCC 40, [2001] 2 S.C.R. 24 at para 3 [*Hudson*] (Can.).

3 Elinor Ostrom, "A Polycentric Approach for Coping with Climate Change," *Annals of Economics and Finance* 15, no. 1 (2014): 106–10.

4 Ostrom, "A Polycentric Approach," 119.

5 Douglas Macdonald, *Carbon Province, Hydro Province: The Challenge of Canadian Energy and Climate Federalism* (Toronto: University of Toronto Press, 2020).

6 *Reference re Greenhouse Gas Pollution Pricing Act*, 2020 ABCA 74 at para 22 [*Reference re GGPPA (ABCA)*] (Can.).

7 Jenna Bednar, William N. Eskridge, Jr., and John Ferejohn, "A Political Theory of Federalism," Working Paper mo. 28763, World Bank Group, Washington DC, 1999, 2, https://documents.worldbank.org/en /publication/documents-reports/documentdetail/900271468761423846 /a-political-theory-of-federalism.

8 In talking about bringing the Green New Deal into the Canadian context, Elizabeth May stated: "I believe that what Canadians really want is peace, order and green government." Stephanie Levitz, "Elizabeth May Seeks to Blow Past Political Pack with Platform Rollout," *National Post*, 16 September 2019, https://nationalpost.com/pmn/news-pmn/canada -news-pmn/greens-to-reveal-platform-and-other-leaders-play-offence -as-campaign-enters-day-6.

9 *Reference re Pan-Canadian Securities Regulation*, 2018 SCC 48, [2018] S.C.R. 189 at para 17 (Can.); *References Re Greenhouse Gas Pollution Pricing Act*, 2021 SCC 11 at para 50.

10 For a discussion of conflict in laws between the federal and provincial governments in the climate context, see Nathalie J. Chalifour, "Jurisdictional Wrangling over Climate Policy in the Canadian Federation: Key Issues in the Provincial Constitutional Challenges to Parliament's Greenhouse Gas Pollution Pricing Act," *Ottawa Law Review* 50, no. 2 (2019): 197–253.

11 *Hudson, supra* note 2 at paras 33, 39–43.

12 *Friends of the Oldman River Society v. Canada (Minister of Transport)*, [1992] 1 S.C.R. 3 at para 64 (Can.).

13 *R. v. Hydro-Québec*, [1997] 3 S.C.R. 213 at para 62 (Can.).

14 For a detailed discussion of the constitutionality of different types of climate measures, See Nathalie J. Chalifour, "Canadian Climate

Federalism: Parliament's Ample Constitutional Authority to Legislate
GHG Emissions through Regulations, a National Cap and Trade
Program, or a National Carbon Tax," *National Journal of Constitutional Law*
36, no. 2 (2016): 331–407.

15 *R v. Hydro-Québec, supra* note 13.

16 *Syncrude Canada Ltd. v. Canada (Attorney General)*, 2016 FCA 160.

17 *References Re Greenhouse Gas Pollution Pricing Act*, 2021 SCC 11.

18 Constitution Act, 1867, 30 & 31 Vict., c 3, s. 91(3) (U.K.), reprinted in
R.S.C. 1985, app II, no 5 (Can.).

19 *Reference re Greenhouse Gas Pollution Pricing Act*, 2019 ONCA 544 at
para 148 [*Reference re GGPPA (ONCA)*] (Can.); *Reference re Greenhouse Gas
Pollution Pricing Act*, 2019 SKCA 40 at paras 84–97 (Can.). See also Sujit
Choudhry, *Constitutional Law and the Politics of Carbon Pricing in Canada*,
IRPP Study no. 74, November 2019, 12.

20 *Reference re GGPPA (SCC), supra* note 17 at para 219; in his dissenting
opinion, Brown J takes issue with most of the majority's core conclusions
but agrees with the majority's characterization of the levies in Parts 1 and
2 of the act as a regulatory charge, not a tax (at para 409).

21 For arguments about the emergency power in the context of climate
change, see Chalifour, "Canadian Climate Federalism," 179–213;
Chalifour, "Jurisdictional Wrangling," 217–53.

22 *Reference re GGPPA (SCC), supra* note 17 at paras 141–2.

23 *Reference re GGPPA (SCC), supra* note 17 at paras 136, 166 (also 141).

24 *Reference re GGPPA (SCC), supra* note 17 at paras 369–70; see Rowe J
adopting Brown J's reasons at para 616.

25 *Reference re GGPPA (SCC), supra* note 17 at paras 167–71.

26 *Reference re GGPPA (SCC), supra* note 17 at paras 369–70; see Rowe J
adopting Brown J's reasons at para 616.

27 See, for example Andrew Leach and Eric M. Adams, "Seeing Double:
Peace, Order, and Good Government, and the Impact of Federal
Greenhouse Gas Emissions Legislation on Provincial Jurisdiction,"
Constitutional Forum Constitutionnel 29, no. 1 (2020): 1–20. See also
Grant Bishop, *Living Tree or Invasive Species? Critical Questions for the
Constitutionality of Federal Carbon Pricing*, Commentary no. 559
(Toronto: C.D. Howe Institute, 2019), https://www.cdhowe.org
/public-policy-research/living-tree-or-invasive-species-critical
-questions-constitutionality-federal-carbon-pricing.

28 Chalifour, "Jurisdictional Wrangling," 37; Martin Olszynski, Nigel Bankes, and Andrew Leach, "Case Comment: Breaking Ranks (and Precedent): Reference re Greenhouse Gas Pollution Pricing Act, 2020 ABCA 74," *Journal of Environmental Law & Practice* 33, no. 2 (2020): 159.

29 For a discussion of the test and how it applies to climate change, see Chalifour, "Jurisdictional Wrangling," 223–5; Sujit Chaudry, "Constitutional Law and the Politics of Carbon Pricing in Canada," Institute for Research in Public Policy Study no. 74, November 2019; Bishop, *Living Tree*; and Jason Maclean, "Climate Change, Constitutions, and Courts: *The Reference re Greenhouse Gas Pollution Pricing Act* and Beyond," *Saskatchewan Law Review* 82 (2019): 147.

30 This approach was also adopted by the majorities in the Ontario and Saskatchewan Courts of Appeal decisions (see note 19). Peter W. Hogg, *Constitutional Law of Canada*, 5th ed. (Scarborough: Thomson Carswell, 2007), **17–15**, quoted in Nathalie J. Chalifour, "Making Federalism Work for Climate Change: Canada's Division of Powers over Carbon Taxes," *National Journal of Constitutional Law* 22, no. 2 (2008): 182.

31 *Reference re GGPPA (ONCA)*, *supra* note 17 at para 20.

32 *Reference re GGPPA (SCC)*, *supra* note 17 at paras 182–92.

33 *Reference re GGPPA (SCC)*, *supra* note 17 at para 191.

34 See, for example, *Reference re GGPPA (ONCA)*, *supra* note 17 at para 106. See also the discussion in Olszynski, Bankes, and Leach "Case Comment," 20.

35 Choudhry "Constitutional Law"; Chalifour, "Jurisdictional Wrangling."

36 *Reference re GGPPA (SCC)*, *supra* note 17 at paras 173–4.

37 *Reference re GGPPA (SCC)*, *supra* note 17 at para 113.

38 *Reference re Pan-Canadian Securities Reference*, 2018 SCC 48.

39 *Reference re GGPPA (SCC)*, *supra* note 17 at para 113.

40 See Choudhry, "Constitutional Law," providing a strong case for climate change as creating a systemic risk.

41 Scott Barrett, *Why Cooperate? The Incentive to Supply Global Public Goods* (Oxford: Oxford University Press, 2007), 47–73.

42 As Huscroft JA notes in dissent at the ONCA, rather than provincial inability "it is, instead, a reflection of legitimate political disagreement on a matter of policy." *Reference re GGPPA (ONCA)*, *supra* note 17 at para 231).

43 Both the majority of the Alberta Court of Appeal and the dissent in the Ontario Court of Appeal also adopted this approach.

44 *Reference re GGPPA (SCC)*, *supra* note 17 at paras 555–7 (Rowe's dissent).

45 *Reference re GGPPA (SCC)*, *supra* note 17 at para 554 (Rowe's dissent).

46 *Reference re GGPPA (SCC)*, *supra* note 17 at para 446 (Brown's dissent).

47 *Reference re GGPPA (SCC)*, *supra* note 17 at para 587.

48 Or as the ABCA put it, with a more malign take on the issue, the federal legislation "is a constitutional Trojan horse" (*Reference re GGPPA (ABCA)*, *supra* note 6 at para 22). Similarly, see Huscroft JA, in dissent, *Reference re GGPPA (ONCA)*, *supra* note 19 at para 227.

49 *Reference re GGPPA (SCC)*, *supra* note 17 at para 387.

50 *Reference re GGPPA (SCC)*, *supra* note 17 at para 388.

51 *Reference re GGPPA (SCC)*, *supra* note 17 at para 196. The Ontario Court of Appeal was willing to find the federal government has struck the appropriate balance given the scale of the climate change problem; see *Reference re GGPPA (ONCA)*, *supra* note 19 at paras 130–6. See also Bishop, "Living Tree, arguing that the federal carbon pricing scheme is not really a national minimum carbon price because of the different industry-specific thresholds.

52 *Reference re GGPPA (SCC)*, *supra* note 17 at paras 199–200.

53 *Reference re GGPPA (SCC)*, *supra* note 17 at paras 205–6.

54 *Reference re GGPPA (SCC)*, *supra* note 17 at paras 2–3; *Reference re GGPPA (ABCA)*, *supra* note 6 at para 11.

55 MacLean, "Climate Change, Constitutions," 193; Chalifour, "Jurisdictional Wrangling," 38–9.

56 *Reference re Securities Act*, 2011 SCC 66 at para 130 [*Securities Act*] (Can.).

57 See Leach and Adams, "Seeing Double," 9–10.

58 *Reference re GGPPA (SCC)*, *supra* note 17 at para 50.

59 *Reference re GGPPA (SCC)*, *supra* note 17 at para 206.

60 Kathryn Harrison, *Passing the Buck: Federalism and Canadian Environmental Policy* (Vancouver: UBC Press, 1996).

61 Political scientist Douglas MacDonald provides a detailed history of climate negotiations in *Carbon Province*, 63–89. See also Douglas MacDonald, Jochen Monstadt, and Kristine Kern, *Allocating Canadian Greenhouse Gas Emission Reductions amongst Sources and Provinces Learning from the European Union, Australia and Germany* (Toronto: University of Toronto, 2013), 49–60, 128–33.

62 MacDonald, Monstadt, and Kern, *Allocating Canadian Greenhouse Gas*, 20, 42.

63 Douglas Macdonald, *Business and Environmental Politics in Canada* (Toronto: University of Toronto Press, 2017); Mark Winfield and Douglas MacDonald, "Federalism and Canadian Climate Change Policy," in *Canadian Federalism: Performance, Effectiveness and Legitimacy*, 3rd ed, ed. Herman Bakvis and Grace Skogstad (Don Mills: Oxford University Press, 2012), quoted in MacDonald, Monstadt, and Kern, *Allocating Canadian Greenhouse Gas*, 47.

64 MacDonald, Monstadt, and Kern, *Allocating Canadian Greenhouse Gas*, 55.

65 MacLean sees two stages to environmental federalism (with the hope for a third): Federalism 1.0 being the old confrontational form while Federalism 2.0 focuses on cooperation and harmonization. He argues that the approach of the Trudeau government does not really represent a new, improved mode, though the government speaks as if it does. Jason MacLean, "Will We Ever Have Paris? Canada's Climate Change Policy and Federalism 3.0," *Alberta Law Review* 55, no. 4 (2018): 889.

66 *House of Commons Debates*, 42–1, vol. 148, no. 299 (23 May 2018), 19550–1 (Right Hon Justin Trudeau), https://www.ourcommons.ca/Content/House/421/Debates/299/HAN299-E.PDF#page=8; see also *House of Commons Debates*, 42–1, vol. 148, no. 293 (7 May 2018), 19189 (Right Hon Justin Trudeau), https://www.ourcommons.ca/Content/House/421/Debates/293/HAN293-E.PDF#page=29.

67 Government of Canada, "Government of Canada Fighting Climate Change with Price on Pollution," news release, 23 October 2018, https://pm.gc.ca/en/news/news-releases/2018/10/23/government-canada-fighting-climate-change-price-pollution.

68 Macdonald, *Carbon Province*, 202–33; MacDonald, *Business and Environmental Politics*; MacLean "Climate Change, Constitutions."

69 Carl Meyer, "Trudeau Goes It Alone with New Climate Plan, Proposes Carbon Price Hike," *Toronto Star*, 11 December 2020, https://www.thestar.com/news/canada/2020/12/11/trudeau-goes-it-alone-with-new-climate-plan-proposes-carbon-price-hike.html.

70 Bednar, Eskridge, and Ferejohn, "Political Theory of Federalism," 2, 14.

71 Jean-François Gaudreault-DesBiens, "Cooperative Federalism in Search of a Normative Justification: Considering the Principle of Federal Loyalty," *Constitutional Forum Constitutionnel* 23, no. 4 (2014): 1–19, https://doi.org/10.21991/C9X68F.

72 Ostrom, "A Polycentric Approach," 121.

73 See Macdonald, *Carbon Province* (pointing to a different aspect of Innis's staples thesis – that the West sees itself as exploited by the centre).
74 Ostrom, "A Polycentric Approach," 108, 124.
75 Macdonald, *Carbon Province*.
76 MacDonald, Monstadt, and Kern, *Allocating Canadian Greenhouse Gas*, 79. See also MacDonald, *Business and Environmental Politics*, 5.
77 Macdonald, *Carbon Province*.
78 See MacDonald, *Business and Environmental Politics*.
79 MacLean, "Will We Ever," 930.
80 *Securities Act, supra* note 56 at para 132. See Jean-Francois Gaudreault-DesBiens, "Cooperative Federalism in Search of a Normative Justification: Considering the Principle of Federal Loyalty," *Constitutional Forum* 23, no. 4 (2014): 1–19, https://doi.org/10.21991/C9X68F (arguing that the court's view of cooperative federalism is too thin and we should consider bringing in an enforceable constitutional principle of federal loyalty as exists in other countries such as Germany).

10 Fostering Trust

1 Jason Maclean, "Striking at the Root Problem of Canadian Environmental Law: Identifying and Escaping Regulatory Capture," *Journal of Environmental Law and Practice* 29 (2016): 113, 125–8.
2 Matthew Hoffmann and Steven Bernstein, "Why Climate Action Gets Stuck – And How We Can Get It Unstuck," TVO, 17 January 2020, https://www.tvo.org/article/why-climate-action-gets-stuck-and-how-we-can-get-it-unstuck; Steven Bernstein and Matthew Hoffmann, "Climate Politics, Metaphors and the Fractal Carbon Trap," *Nature Climate Change* 9 (2019): 919–25.
3 See, for example, David R. Boyd, *Unnatural Law: Rethinking Canadian Environmental Law and Policy* (Vancouver: UBC Press, 2003), 292–3; Bruce Pardy, "In Search of the Holy Grail of Environmental Law: A Rule to Solve the Problem," *McGill International Journal of Sustainable Development Law* 1 (2005): 36–7.
4 *Climate Change Accountability Act*, S.B.C. 2007, c. 42, s. 2(1) (Can.).
5 *Oil Sands Emissions Limit Act*, S.A. 2016, c. O-7.5, s. 2(1) (Can.).
6 *Canadian Net-Zero Emissions Accountability Act*, S.C. 2021, c. 22.

7 *Climate Change Act 2008*, c. 27, ss. 1(1), 4 (UK).

8 Andrew Green, "On Thin Ice: Meeting Canada's Paris Climate Commitments," *Journal of Environmental Law and Practice* 32, no. 1 (2018): 122, 111–12; Canadian Institute for Climate Choices, *Climate Legislation in the United Kingdom* (Ottawa: CICC, March 2020), 7, https:// climatechoices.ca/wp-content/uploads/2020/03/CICC-Climate -Legislation-in-the-United-Kingdom.pdf.

9 Cameron Hepburn, Stephen Duncan, and Antonis Papachristodoulou, "Behavioural Economics, Hyperbolic Discounting and Environmental Policy," *Environmental and Resource Economics* 46, no. 2 (June 2010): 192.

10 See, for example, Cass R. Sunstein, "On the Expressive Function of Law," *University of Pennsylvania Law Review* 144, no. 5 (1996): 2024–5.

11 *Impact Assessment Act*, S.C. 2019, c. 28, s. 1, s. 63 (Can.).

12 Climate Change Act 2008, c. 27, s. 10(2) (UK).

13 *Environment Quality Act*, C.Q.L.R., c. Q-2, s. 46.4 (Can.).

14 *Canadian Net-Zero Emissions Accountability Act*, s. 8.

15 Jocelyn Stacey, "The Environmental Emergency and the Legality of Discretion in Environmental Law," *Osgoode Hall Law Journal* 52, no. 3 (2015): 989, 1002, 1010–13; Pardy, "Holy Grail," 32–3.

16 Amartya Sen, *Development as Freedom* (New York: Anchor Books, 2000).

17 See William D. Nordhaus and James Tobin, "Is Growth Obsolete?" in *Economic Research: Retrospect and Prospect*, vol. 5: *Economic Growth*, ed. William D. Nordhaus and James Tobin (New York: NBER, 1972), 4; Joseph E. Stiglitz, Amartya Sen, and Jean-Paul Fitoussi, *Mismeasuring Our Lives: Why GDP Doesn't Add Up* (New York: New Press, 2010). For a summary of substitutes for GDP, see L. Fasolo, M. Galetto, and E. Turina, "A Pragmatic Approach to Evaluate Alternative Indicators to GDP," *Quality and Quantity* 47, no. 2 (2013): 635–8.

18 Michael Anderson and Elias Mossialos, "Beyond Gross Domestic Product for New Zealand's Wellbeing Budget," *Lancet* 4, no. 7 (2019): e320–1; see also Christoph Schumacher, "New Zealand's 'Well-Being Budget': How It Hopes to Improve People's Lives," *The Conversation*, 30 May 2019, https://theconversation.com/new-zealands-well-being-budget-how-it -hopes-to-improve-peoples-lives-118052.

19 Boyd, *Unnatural Law*.

20 Sustainable Development Goals Act, S.N.S. 2019, c. 26, s. 4(a) (Can.).

21 See, for example, Environmental Bill of Rights, 1993, S.O. 1993, c. 28, ss. 2(3)(a), 8 (Can.), and the federal Impact Assessment Act, S.C. 2019, c. 28, s. 1, s. 11 (Can.)

22 Mark Winfield, *A New Era of Environmental Governance in Canada: Better Decisions Regarding Infrastructure and Resource Development Projects*, Green Prosperity Papers (Toronto: Metcalf Foundation, 2016), 7–8, https:// metcalffoundation.com/site/uploads/2016/05/Metcalf_Green -Prosperity-Papers_Era-of-Governance_final_web.pdf.

23 Jason MacLean, "Will We Ever Have Paris? Canada's Climate Change Policy and Federalism 3.0," *Alberta Law Review* 55, no. 4 (2018): 930.

24 There are, of course, differences. For the pandemic, the public might have been willing to listen to experts because of the immediate fear of death or grave illness. Moreover, the short time frame might also make the public more willing to attend to expertise than in the case of decision-making spread over decades.

25 Mark Jaccard, Mikela Hein, and Tiffany Vass, *Is Win-Win Possible? Can Canada's Government Achieve Its Paris Commitment ... and Get Re-elected?* (Vancouver: School of Resource and Environmental Management, Simon Fraser University, 2016), 8, http://www.sfu.ca/content/dam/sfu/emrg /Publications/Research_Publications/2016%20Jaccard,%20Hein%20& %20Vass%20-%20Is%20Win-Win%20Possible.pdf; Canadian Institute for Climate Choices, *United Kingdom*.

26 Species at Risk Act, S.C. 2002, c. 29.

27 Climate Change Act 2008, c. 27, sched. 1 (UK).

28 Canadian Institute for Climate Choices, *United Kingdom*, 7, 9.

29 Ron Ellis, *Unjust by Design: Canada's Administrative Justice System* (Vancouver: UBC Press, 2013).

30 Stacey, "Environmental Emergency," 1015.

31 Climate Change Accountability Act, S.B.C. 2007, c. 42, s. 4.2 (Can.). See also Manitoba's The Climate and Green Plan Act, C.C.S.M. c. C134, ss 2(2), 7(1) (Can.).

32 Canadian Net-Zero Emissions Accountability Act, s. 21(1.1).

33 Canadian Net-Zero Emissions Accountability Act, s. 24.

34 Boyd, *Unnatural Law*, 293–4; David R. Boyd, *The Right to a Healthy Environment: Revitalizing Canada's Constitution* (Vancouver: UBC Press, 2012), 3; Lynda M. Collins, "Safeguarding the Longue Durée: Environmental Rights in the Canadian Constitution," *Supreme Court Law*

Review 71 (2015): paras 35–8; Lynda M. Collins and David R. Boyd, "Non-regression and the Charter Right to a Healthy Environment," *Journal of Environmental Law and Practice* 29 (2016): 286–7; Lynda Collins and Lorne Sossin, "In Search of an Ecological Approach to Constitutional Principles and Environmental Discretion in Canada," *UBC Law Review* 52, no. 1 (2019): 295.

35 Collins and Boyd, "Non-regression," 287, 293.
36 Jason MacLean, "Climate Change, Constitutions, and Courts: The *Reference re Greenhouse Gas Pollution Pricing Act* and Beyond," *Saskatchewan Law Review* 82, no. 2 (2019): 178–9, 182–3.
37 Collins and Sossin, "Ecological Approach," 294.
38 *Reference re Secession of Quebec*, [1998] 2 S.C.R. 217 at paras 32, 49 (Can.).
39 Collins and Sossin, "Ecological Approach," 342.
40 *Toronto (City) v. Ontario (AG)*, 2021 SCC 34.
41 See, for example, Environmental Bill of Rights, 1993, S.O. 1993, c. 28 (Can.).
42 Bill C-28, An Act to amend the Canadian Environmental Protection Act, 1999.
43 *United Nations Declaration on the Rights of Indigenous Peoples Act*, S.C. 2021, c. 14.
44 Naiomi Metallic, "Deference and Legal Frameworks Not Designed By, For or With Us," *Administrative Law Matters* (blog), 27 February 2018, https://www.administrativelawmatters.com/blog/2018/02/27/deference-and-legal-frameworks-not-designed-by-for-or-with-us-naiomi-metallic/.
45 Andrew Green, "Judicial Influence on the Duty to Consult and Accommodate," *Osgoode Hall Law Journal* 56, no. 3 (2020): 529.
46 Jason MacLean and Chris Tollefson, "Climate-Proofing Judicial Review after Paris: Judicial Competence, Capacity, and Courage," *Journal of Environmental Law and Practice* 31, no. 3 (2018): 245–70, 252.
47 Adrian Vermeule, *Law's Abnegation: From Law's Empire to the Administrative State* (Cambridge, MA: Harvard University Press, 2016), 126.
48 MacLean and Tollefson, "Climate-Proofing," 270.
49 Stacey, "Environmental Emergency," 989, quoting David Dyzenhaus, "Law as Justification: Etienne Mureinik's Conception of Legal Culture," *South African Journal on Human Rights* 14, no. 1 (1998): 30.
50 Stacey, "Environmental Emergency," 1023 (discussing the environmental assessment process).

51 Bruce Pardy, "The Unbearable Licence of Being the Executive: A Response to Stacey's Permanent Environmental Emergency," *Osgoode Hall Law Journal* 52, no. 3 (2015): 1043–4.
52 *Canada (Minister of Citizenship and Immigration) v. Vavilov*, 2019 SCC 65 at paras 2, 82–138.
53 *References re Greenhouse Gas Pollution Pricing Act*, 2021 SCC 11, at 73.
54 *Manitoba v. Canada (Attorney General)*, 2021 FC 1115.
55 See Cass R. Sunstein, "*Chevron* as Law," *Georgetown Law Journal*, forthcoming.
56 *Friends of the Earth v. Canada (Environment)*, 2009 FCA 297.
57 Neil Komesar, *Imperfect Alternatives: Choosing Institutions in Law, Economics and Public Policy* (Chicago: University of Chicago Press, 1997).

11 Setting the Foundation

1 Mark Jaccard, *The Citizen's Guide to Climate Success: Overcoming Myths That Hinder Progress* (Cambridge: Cambridge University Press, 2020).
2 Thomas Friedman, "A Scary Energy Winter is Coming. Don't Blame the Greens," *New York Times*, 5 October 2021.
3 Jennie Wang and Abdoul-R. Mamane, "Household Food Consumption and Canadian Greenhouse Gas Emissions, 2015," Statistics Canada, 9 October 2019, https://www150.statcan.gc.ca/n1/pub/16-508-x/16-508-x2019004-eng.htm.
4 Clark Milito and Gabriel Gagnon, "Greenhouse Gas Emissions – A Focus on Canadian Households," Statistics Canada, 2008), https://www150.statcan.gc.ca/n1/pub/16-002-x/2008004/article/10749-eng.htm.
5 Wang and Mamane, "Household Food Consumption."
6 ECCC, *Pan-Canadian Framework on Clean Growth and Climate Change: Third Annual Synthesis Report on the Status of Implementation* (Gatineau: ECCC, 2019), 13, http://publications.gc.ca/collections/collection_2020/eccc/En1-77-2019-eng.pdf.
7 Cherise Burda et al., *Behind the Wheel: Opportunities for Canadians to Drive Less, Reduce Pollution and Save Money* (Drayton Valley: Pembina Institute, 2012), 2, https://www.pembina.org/reports/behind-the-wheel.pdf.
8 Government of Canada, *National Inventory Report 1990–2017*, 5, table A9–2; Government of Canada, *National Inventory Report 1990–2015*, 82, table 2–12.

9 Burda et al., *Behind the Wheel*, 5, table 1.

10 Natural Resources Canada, "Canada's GHG Emissions by Sector, End Use and Subsector – Including Electricity-Related Emissions," accessed 7 February 2022, https://oee.nrcan.gc.ca/corporate/statistics/neud /dpa/showTable.cfm?type=HB&sector=aaa&juris=ca&rn=3 &page=0.

11 Government of Canada, *Canada's 7th National Communication and 3rd Biennial Report* (Gatineau: ECCC, 2017), 145, http://publications.gc.ca /collections/collection_2018/eccc/En4-73-2017-eng.pdf.

12 Canada Energy Regulator, "Market Snapshot: Greenhouse Gas Emissions Associated with Residential Electricity Consumption Vary Significantly by Province and Territory," 21 June 2017, https://www.cer-rec.gc.ca /nrg/ntgrtd/mrkt/snpsht/2017/06-04grngsmssnsrsdntl-eng.html.

13 Canada Energy Regulator, "Market Snapshot."

14 Environmental Commissioner of Ontario, *Facing Climate Change: Greenhouse Gas Progress Report 2016* (Toronto: ECO, 2016), 55–6, https:// media.assets.eco.on.ca/web/2016/11/2016-Annual-GHG-Report _Chapter-3.pdf.

15 Sarah Dobson and G. Kent Fellows, "Big and Little Feet: A Comparison of Provincial Level Consumption and Production-Based Emissions Footprints," University of Calgary School of Public Policy Publications 10, no. 23 (2017).

16 Environmental Commissioner of Ontario, *Facing Climate Change*, 55. See also Craig Gaston, "Consumption-Related Greenhouse Gas Emissions in Canada, the United States and China," *Envirostats* 5, no. 4 (2011): 14–21; Brett Dolter and Peter A. Victor, "Casting a Long Shadow: Demand-Based Accounting of Canada's Greenhouse Gas Emissions Responsibility," *Ecological Economics* 127 (2016): 156–64.

17 Wang and Mamane, "Household Food Consumption."

18 For a good discussion of where the main reductions come from, see Saul Griffith, "How to Solve Climate Change and Make Life More Awesome," interview by Ezra Klein, 16 December 2019, *The Ezra Klein Show* (podcast), https://www.stitcher.com/podcast/the-ezra-klein-show/e /66028720?autoplay=true.

19 Thomas Deitz et al., "Household Actions Can Provide a Behavioral Wedge to Rapidly Reduce U.S. Carbon Emissions," *Proceedings of the National Academy of Sciences* 106, no. 4 (2009): 18453, table 1.

20 Naomi Klein, *On Fire: The Burning Case for a Green New Deal* (Toronto: Simon and Schuster, 2019), 262.
21 Jaccard, *Citizen's Guide*, 154, 159–60.
22 Environics Institute for Survey Research, *Focus Canada – Fall 2019: Regional Perspectives on Politics and Priorities* (Toronto: EISR, 2019), 4, https://www.environicsinstitute.org/docs/default-source/default -document-library/focus-canada-fall-2019-survey-_report-2-priorities -final.pdf?sfvrsn=42e05492_0.
23 Positive Energy, "COVID-19 Puts Canadians at the Fulcrum of Environment and Economy," https://www.uottawa.ca/positive-energy /content/covid-19-puts-canadians-fulcrum-environment-and-economy.
24 Nanos, "Slim Majority Show Some Willingness to Pay More to Help Achieve Canada's Emission Reduction Targets," September 2021, https://nanos.co/wp-content/uploads/2021/09/2021-1959-1960-CTV -Globe-August-Climate-Change-Populated-Report-with-tabs.pdf.
25 Nanos, "Slim Majority."
26 Richard H. MacAdams, "The Origin, Development and Regulation of Norms," *Michigan Law Review* 96, no. 2 (1997): 340, 355–81; Eric A. Posner, *Law and Social Norms* (Cambridge, MA: Harvard University Press, 2002).
27 See George A. Akerlof and Rachel E. Kranton, *Identity Economics: How Our Identities Shape Our Work, Wages, and Well-Being* (Princeton: Princeton University Press, 2010); Andrew Green, "Creating Environmentalists: Environmental Law, Identity and Commitment," *Journal of Environmental Law and Practice* 17, no. 1 (2006): 1–26.
28 More generally, see Cass R. Sunstein and Lucia A. Reisch, "Greener by Default," Discussion Paper no. 951, John M. Olin Discussion Paper Series, Harvard Law School, Cambridge, MA, 2018, http://www.law.harvard .edu/programs/olin_center/papers/pdf/Sunstein_951.pdf.
29 John M. Gowdy, "Behavioural Economics and Climate Change Policy," *Journal of Economic Behaviour and Organization* 68, no. 3–4 (2008): 635.
30 Uri Gneezy and Aldo Rustichini, "A Fine Is a Price," *Journal of Legal Studies* 29, no. 1 (2000): 1–17. For more information on the utilization and effectiveness of incentive mechanisms, see Uri Gneezy and Aldo Rustichini, "Pay Enough or Don't Pay At All," *Quarterly Journal of Economics* 115, no. 3 (2000): 791–810; Uri Gneezy, Stephan Meier, and Pedro Rey-Biel, "When and Why Incentives (Don't) Work to Modify Behaviour," *Journal of Economic Perspectives* 25, no. 4 (2011): 191–210.

31 Cherie Metcalf, Emily Satterthwaite, Shahar Dillbary, and Brock
    Stoddard, "Is a Fine Still a Price? Replication as Robustness in Empirical
    Legal Studies" *International Review of Law and Economics* 63 (2020): 1.
32 Gowdy, "Behavioral Economics."
33 Saurabh Bhargava and George Loewenstein, "Behavioural Economics
    and Public Policy 102: Beyond Nudging," *American Economic Review* 105,
    no. 5 (2015): 396–401.
34 Jaccard, *Citizen's Guide*, 34.
35 See, for example, Dan M. Kahan, "What Is the 'Science of Science
    Communication'?" *Journal of Science Communication* 14, no. 3 (2015):
    1824–2049.
36 There is a significant literature on individual choices and climate change.
    See, for example, Michael P. Vandenbergh, Jack Barkenbus, and Jonathan
    Gilligan, "Individual Carbon Emissions: The Low-Hanging Fruit," *UCLA
    Law Review* 55, no. 6 (2007–08): 1701–58.
37 Robert Gifford, Christine Kormos, and Amanda McIntyre, "Behavioral
    Dimensions of Climate Change: Drivers, Responses, Barriers, and
    Interventions," *Wiley Interdisciplinary Reviews: Climate Change* 2, no. 6
    (2011): 818; Davide Quaglione et al., "An Assessment of the Role of
    Cultural Capital on Sustainable Mobility Behaviours: Conceptual
    Framework and Empirical Evidence," *Socio-Economics Planning Sciences*
    66 (2018): 24–34. More generally, see Martin Kesternich, Christiane
    Reif, and Dirk Rübbelke, "Recent Trends in Behavioral Environmental
    Economics," *Environmental and Resource Economics* 67, no. 3 (2017): 403–11.
38 Government of Canada, *Canada's Green Plan: Canada's Green Plan for
    a Healthy Environment* (Ottawa: Supply and Services Canada, 1990),
    9, http://cfs.nrcan.gc.ca/pubwarehouse/pdfs/24604.pdf; Heather
    A. Smith, "Political Parties and Canadian Climate Change Policy,"
    International Journal 64, no. 1 (2008/2009): 49.
39 Mark Jaccard, Mikela Hein, and Tiffany Vass, *Is Win-Win Possible? Can
    Canada's Government Achieve Its Paris Commitment … and Get Re-elected?*
    (Vancouver: School of Resource and Environmental Management, Simon
    Fraser University, 2016), 5, http://www.sfu.ca/content/dam/sfu/emrg
    /Publications/Research_Publications/2016%20Jaccard,%20Hein%20
    &%20Vass%20-%20Is%20Win-Win%20Possible.pdf.
40 Alexa Spence and Nick Pidgeon, "Framing and Communicating Climate
    Change: The Effects of Distance and Outcome Frame Manipulations,"

Global Environmental Change 20, no. 4 (2010): 656–67; Omar I. Asensio and Magali A. Delmas, "The Dynamics of Behavior Change: Evidence from Energy Conservation," *Journal of Economic Behavior & Organization* 126 (2016): 196–212.

41 David Hagmann, Emily H. Ho, and George Loewenstein, "Nudging Out Support for a Carbon Tax," *Nature Climate Change* 9 (2019): 484; Jon M. Jachimowicz et al., "The Critical Role of Second-Order Normative Beliefs in Predicting Energy Conservation," *Nature Human Behaviour* 2, no. 10 (2018): 757–64.

42 Cass R. Sunstein and Lucia A. Reisch, "Climate-Friendly Default Rules," Discussion Paper no. 878, John M. Olin Discussion Paper Series, Harvard Law School, Cambridge, MA, 2016, 4, http://www.law.harvard.edu /programs/olin_center/papers/pdf/Sunstein_878.pdf.

43 Sunstein and Reisch, "Climate-Friendly Default Rules," 5.

44 Zachary Brown et al., "Testing the Effect of Defaults on the Thermostat Settings of OECD Employees," *Energy Economics* 39 (2013): 128–34.

45 Sunstein and Reisch, "Climate-Friendly Default Rules."

46 Saurabh Bhargava and George Lowenstein, "Behavioral Economics and Public Policy 102: Beyond Nudging," *American Economic Review* 105, no. 5 (2015), 397.

47 Cass R. Sunstein and Lucia A. Reisch, "Automatically Green: Behavioral Economics and Environmental Protection," *Harvard Environmental Law Review* 38, no. 1 (2014): 157–8.

48 See, for example, Michael G. Pollitt and Irina Shaorshadze, "The Role of Behavioural Economics in Energy and Climate Policy," EPRG Working Paper no. 1130, Electricity Policy Research Group, University of Cambridge, 2011, https://doi.org/10.17863/CAM.5237.

49 Allison Jones, "Sales of Electric Vehicles Drop Sharply in Ontario after Rebate Cancellation," *Globe and Mail*, 15 December 2019, https://www .theglobeandmail.com/business/article-sales-of-electric-vehicles -plummet-in-ontario-after-rebate/.

50 Jones, "Sales."

51 Elizabeth Beale et al., *Supporting Carbon Pricing: How to Identify Policies that Genuinely Complement an Economy-Wide Carbon Price* (Montreal: Canada's Ecofiscal Commission, 2017), 42–8, 129, http://ecofiscal .ca/wp-content/uploads/2017/06/Ecofiscal-Commission-Report -Supporting-Carbon-Pricing-June-2017.pdf.

52 Ghislain Dubios et al., "It Starts at Home? Climate Policies Targeting Household Consumption and Behavioral Decisions Are Key to Low-Carbon Futures," *Energy Research & Social Science* 52 (2019): 148.
53 Cass Sunstein, "On the Expressive Function of Law," *University of Pennsylvania Law Review* 14 (1996): 2021.
54 Beale et al., *Supporting Carbon Pricing*, 46; Sunstein and Reisch, "Automatically Green," 129.
55 Cass R. Sunstein, "Switching the Default Rule," *New York University Law Review* 77, no. 1 (2002): 111.
56 Sunstein and Reisch, "Climate-Friendly Default Rules," 19.
57 Sunstein and Reisch, "Climate-Friendly Default Rules," 19.
58 Amartya Sen, *Development as Freedom* (New York: Oxford University Press, 1999), 276, 38.
59 See, for example, the work of Shaun Fluker on public participation in environmental decisions, such as "The Right to Public Participation in Resource and Environmental Decision-Making in Alberta," *Alberta Law Review* 52, no. 3 (2015): 567.
60 Andrew Green, "Delegation and Consultation: How the Administrative State Functions," in *Administrative Law in Context*, 3rd ed., ed. Colleen Flood and Lorne Sossin (Toronto: Emond, 2017).
61 Howard Kunreuther, "What the COVID-19 Curve Can Teach Us about Climate Change," *Knowledge@Wharton*, Wharton School, University of Pennsylvania, 7 April 2020, https://knowledge.wharton.upenn.edu/article/what-can-the-covid-19-pandemic-teach-us-about-climate-change/.
62 Bill Weir, "What Coronavirus Could Teach Us about Climate Change" (video), CNN, 2020, www.cnn.com/videos/weather/2020/03/26/weir-climate-crisis-impact-coronavirus-project-planet-orig.cnn/video/playlists/project-planet/.

12 Breaking the Cycle

1 Canadian Institute for Climate Choices, *Canada's Net Zero Future: Finding Our Way in the Global Transition* (Ottawa: CICC, 2021), https://climatechoices.ca/reports/canadas-net-zero-future/.
2 Mark Jaccard, *The Citizen's Guide to Climate Success: Overcoming Myths That Hinder Progress* (Cambridge: Cambridge University Press, 2020), 234 (calls this a "Faustian dilemma").

3 See, for example, Brendan Haley, "The Staple Theory and the Carbon Trap," in *The Staple Theory @ 50: Reflections on the Lasting Significance of Mel Watkins "A Staple Theory of Economic Growth,"* ed. Jim Stanford (Canadian Centre for Policy Alternatives, 2014), 77; Jason MacLean, "Striking at the Root Problem of Canadian Environmental Law: Identifying and Escaping Regulatory Capture," *Journal of Environmental Law and Practice* 29 (2016): 111–28.

4 David Boyd, *Unnatural Law: Rethinking Canadian Environmental Law and Policy* (Vancouver: UBC Press, 2003).

5 Stephen Breyer, *Breaking the Vicious Cycle: Toward Effective Risk Regulation* (Cambridge, MA: Harvard University Press, 1995).

6 Steven Bernstein and Matthew Hoffman, "Climate Politics, Metaphors and the Fractal Carbon Trap," *Nature Climate Change* 9 (2019): 919–25, https://doi.org/10.1038/s41558-019-0618-2.

7 Jeff Colgan, Jessica Green, and Thomas Hale, "Asset Revaluation and the Existential Politics of Climate Change," *International Organization*, forthcoming.

8 Albert O. Hirschman, "Against Parsimony: Three Easy Ways of Complicating Some Categories of Economic Discourse," *American Economic Review* 74, no. 2 (1984): 90, http://www.jstor.org/stable/1816336.

9 Hirschman, "Against Parsimony," 93.

Index

Page numbers in *italics* refer to figures

in **UTP** insights

Books in the Series

- Michael R. Marrus, *Lessons of the Holocaust*
- Roland Paris and Taylor Owen (eds.), *The World Won't Wait: Why Canada Needs to Rethink Its International Policies*
- Bessma Momani, *Arab Dawn: Arab Youth and the Demographic Dividend They Will Bring*
- William Watson, *The Inequality Trap: Fighting Capitalism Instead of Poverty*
- Phil Ryan, *After the New Atheist Debate*
- Paul Evans, *Engaging China: Myth, Aspiration, and Strategy in Canadian Policy from Trudeau to Harper*

www.ingramcontent.com/pod-product-compliance
Lightning Source LLC
Chambersburg PA
CBHW030237030426
42336CB00009B/145